RELATIVE INVARIANTS OF RINGS

PURE AND APPLIED MATHEMATICS

A Program of Monographs, Textbooks, and Lecture Notes

MONOGRAPHS AND TEXTBOOKS IN PURE AND APPLIED MATHEMATICS

1. *K. Yano,* Integral Formulas in Riemannian Geometry (1970) *(out of print)*
2. *S. Kobayashi,* Hyperbolic Manifolds and Holomorphic Mappings (1970) *(out of print)*
3. *V. S. Vladimirov,* Equations of Mathematical Physics (A. Jeffrey, editor; A. Littlewood, translator) (1970) *(out of print)*
4. *B. N. Pshenichnyi,* Necessary Conditions for an Extremum (L. Neustadt, translation editor; K. Makowski, translator) (1971)
5. *L. Narici, E. Beckenstein, and G. Bachman,* Functional Analysis and Valuation Theory (1971)
6. *D. S. Passman,* Infinite Group Rings (1971)
7. *L. Dornhoff,* Group Representation Theory (in two parts). Part A: Ordinary Representation Theory. Part B: Modular Representation Theory (1971, 1972)
8. *W. Boothby and G. L. Weiss (eds.),* Symmetric Spaces: Short Courses Presented at Washington University (1972)
9. *Y. Matsushima,* Differentiable Manifolds (E. T. Kobayashi, translator) (1972)
10. *L. E. Ward, Jr.,* Topology: An Outline for a First Course (1972) *(out of print)*
11. *A. Babakhanian,* Cohomological Methods in Group Theory (1972)
12. *R. Gilmer,* Multiplicative Ideal Theory (1972)
13. *J. Yeh,* Stochastic Processes and the Wiener Integral (1973) *(out of print)*
14. *J. Barros-Neto,* Introduction to the Theory of Distributions (1973) *(out of print)*
15. *R. Larsen,* Functional Analysis: An Introduction (1973) *(out of print)*
16. *K. Yano and S. Ishihara,* Tangent and Cotangent Bundles: Differential Geometry (1973) *(out of print)*
17. *C. Procesi,* Rings with Polynomial Identities (1973)
18. *R. Hermann,* Geometry, Physics, and Systems (1973)
19. *N. R. Wallach,* Harmonic Analysis on Homogeneous Spaces (1973) *(out of print)*
20. *J. Dieudonné,* Introduction to the Theory of Formal Groups (1973)
21. *I. Vaisman,* Cohomology and Differential Forms (1973)
22. *B.-Y. Chen,* Geometry of Submanifolds (1973)
23. *M. Marcus,* Finite Dimensional Multilinear Algebra (in two parts) (1973, 1975)
24. *R. Larsen,* Banach Algebras: An Introduction (1973)
25. *R. O. Kujala and A. L. Vitter (eds.),* Value Distribution Theory: Part A; Part B. Deficit and Bezout Estimates by Wilhelm Stoll (1973)
26. *K. B. Stolarsky,* Algebraic Numbers and Diophantine Approximation (1974)
27. *A. R. Magid,* The Separable Galois Theory of Commutative Rings (1974)
28. *B. R. McDonald,* Finite Rings with Identity (1974)
29. *J. Satake,* Linear Algebra (S. Koh, T. Akiba, and S. Ihara, translators) (1975)

30. *J. S. Golan,* Localization of Noncommutative Rings (1975)
31. *G. Klambauer,* Mathematical Analysis (1975)
32. *M. K. Agoston,* Algebraic Topology: A First Course (1976)
33. *K. R. Goodearl,* Ring Theory: Nonsingular Rings and Modules (1976)
34. *L. E. Mansfield,* Linear Algebra with Geometric Applications: Selected Topics (1976)
35. *N. J. Pullman,* Matrix Theory and Its Applications (1976)
36. *B. R. McDonald,* Geometric Algebra Over Local Rings (1976)
37. *C. W. Groetsch,* Generalized Inverses of Linear Operators: Representation and Approximation (1977)
38. *J. E. Kuczkowski and J. L. Gersting,* Abstract Algebra: A First Look (1977)
39. *C. O. Christenson and W. L. Voxman,* Aspects of Topology (1977)
40. *M. Nagata,* Field Theory (1977)
41. *R. L. Long,* Algebraic Number Theory (1977)
42. *W. F. Pfeffer,* Integrals and Measures (1977)
43. *R. L. Wheeden and A. Zygmund,* Measure and Integral: An Introduction to Real Analysis (1977)
44. *J. H. Curtiss,* Introduction to Functions of a Complex Variable (1978)
45. *K. Hrbacek and T. Jech,* Introduction to Set Theory (1978) *(out of print)*
46. *W. S. Massey,* Homology and Cohomology Theory (1978)
47. *M. Marcus,* Introduction to Modern Algebra (1978)
48. *E. C. Young,* Vector and Tensor Analysis (1978)
49. *S. B. Nadler, Jr.,* Hyperspaces of Sets (1978)
50. *S. K. Sehgal,* Topics in Group Rings (1978)
51. *A. C. M. van Rooij,* Non-Archimedean Functional Analysis (1978)
52. *L. Corwin and R. Szczarba,* Calculus in Vector Spaces (1979)
53. *C. Sadosky,* Interpolation of Operators and Singular Integrals: An Introduction to Harmonic Analysis (1979)
54. *J. Cronin,* Differential Equations: Introduction and Quantitative Theory (1980)
55. *C. W. Groetsch,* Elements of Applicable Functional Analysis (1980)
56. *I. Vaisman,* Foundations of Three-Dimensional Euclidean Geometry (1980)
57. *H. I. Freedman,* Deterministic Mathematical Models in Population Ecology (1980)
58. *S. B. Chae,* Lebesgue Integration (1980)
59. *C. S. Rees, S. M. Shah, and Č. V. Stanojević,* Theory and Applications of Fourier Analysis (1981)
60. *L. Nachbin,* Introduction to Functional Analysis: Banach Spaces and Differential Calculus (R. M. Aron, translator) (1981)
61. *G. Orzech and M. Orzech,* Plane Algebraic Curves: An Introduction Via Valuations (1981)
62. *R. Johnsonbaugh and W. E. Pfaffenberger,* Foundations of Mathematical Analysis (1981)

63. *W. L. Voxman and R. H. Goetschel,* Advanced Calculus: An Introduction to Modern Analysis (1981)
64. *L. J. Corwin and R. H. Szczarba,* Multivariable Calculus (1982)
65. *V. I. Istrǎtescu,* Introduction to Linear Operator Theory (1981)
66. *R. D. Järvinen,* Finite and Infinite Dimensional Linear Spaces: A Comparative Study in Algebraic and Analytic Settings (1981)
67. *J. K. Beem and P. E. Ehrlich,* Global Lorentzian Geometry (1981)
68. *D. L. Armacost,* The Structure of Locally Compact Abelian Groups (1981)
69. *J. W. Brewer and M. K. Smith, eds.,* Emmy Noether: A Tribute to Her Life and Work (1981)
70. *K. H. Kim,* Boolean Matrix Theory and Applications (1982)
71. *T. W. Wieting,* The Mathematical Theory of Chromatic Plane Ornaments (1982)
72. *D. B. Gauld,* Differential Topology: An Introduction (1982)
73. *R. L. Faber,* Foundations of Euclidean and Non-Euclidean Geometry (1983)
74. *M. Carmeli,* Statistical Theory and Random Matrices (1983)
75. *J. H. Carruth, J. A. Hildebrant, and R. J. Koch,* The Theory of Topological Semigroups (1983)
76. *R. L. Faber,* Differential Geometry and Relativity Theory: An Introduction (1983)
77. *Stephen Barnett,* Polynomials and Linear Control Systems (1983)
78. *Gregory Karpilovsky,* Commutative Group Algebras (1983)
79. *F. Van Oystaeyen and A. Verschoren,* Relative Invariants of Rings: The Commutative Theory (1983)

Other Volumes in Preparation

RELATIVE INVARIANTS OF RINGS

The Commutative Theory

F. Van Oystaeyen • A. Verschoren
Department of Mathematics
University of Antwerp
Wilrijk, Belgium

MARCEL DEKKER, INC. New York and Basel

Library of Congress Cataloging in Publication Data

Oystaeyen, F. van
Relative invariants of rings.

(Monographs and textbooks in pure and applied mathematics ; v. 79)
Bibliography: p.
Includes indexes.
1. Commutative rings. 2. Invariants. I. Verschoren, A. . II. Title. III. Series.
QA251.3.095 1983 512'.4 83-14401
ISBN 0-8247-7043-9

MARCEL DEKKER, INC.
270 Madison Avenue, New York, New York 10016

Current printing (last digit):
10 9 8 7 6 5 4 3 2 1

PRINTED IN THE UNITED STATES OF AMERICA

To the memory of Florintine and Jozef Vervynck
F. Van Oystaeyen

To Lea and Hedwig
A. Verschoren

PREFACE

Associating groups to mathematical objects in order to classify or characterize these seems to be a very natural mathematical action. Obvious examples that press forward are the fundamental groups of topological spaces, the homotopy, homology or cohomology groups, the K-groups etc... . In this book we investigate certain groups associated to associative rings, in particular : Picard groups, class groups, Brauer groups or related cohomology groups. These groups are determined by the isomorphism class of the ring considered and they may be considered to be "invariants" of the ring structure. The main aim is to infer knowledge about the structure of the ring from knowledge about certain invariants. It seldom happens that knowledge about one or a few invariants allows to characterize the ring, so it is usually necessary to consider more complicated invariants in order to get closer to our goal of obtaining useful structure results for the rings studied.

The so-called "relative invariants" we introduce in Part I and Part II of this work are invariants which are parametrized by certain localizations of some commutative ground ring. Picard groups and certain types of class groups as well as the relative versions of these may be defined for general non-commutative rings, in particular for orders in central simple algebras. It is to be expected that a theory of relative invariants for orders, over Dedekind or Krull domains, in a central simple algebra will be linked to the arithmetic of the central simple algebra and the arithmetical ideal theory of the order considered. Actually the introduction of the relative invariants of orders introduces an aspect of arithmetical module theory in this theory because the new invariants are not always given in terms of ideals. All this is the topic of Part II and none of it appears in this book which is entirely devoted to relative invariants of commutative rings. There are at least two very good reasons for this division. Firstly, the Brauer group and its relative versions are only defined in the commutative case. Secondly, the relative invariants are related to the scheme Spec(R), so it is not surprising to see that they obtain a geometrical meaning which accounts for some applications of pure Algebraic Geometrical natúre. In geometrical terminology the relative invariants we introduce may be viewed as sections over a geometrically stable (e.g. an open) subspace of, the Brauer group or the Picard group say, of the ringed space Spec(R). So, after the technical Chapters II and III, we include an extensive treatment of the sheaf theoretical methods involved i.e. Brauer groups and Picard groups of ringed spaces and quasi-affine schemes. The relative theory may then be applied to yield some interesting results on the extension of coherent and quasicoherent sheaves from geometrically stable subspaces of Spec(R) to Spec(R), or yield new proofs of some well-known results of G. Horrocks.

In Chapter V we give some ring theoretical applications; in particular certain parts of V.2. have some importance in the light of the proposed treatment of orders in Part II. In order to obtain a projective counterpart for the duality between relative invariants and invariants of ringed spaces, one has to study relative invariants of commutative $\mathbb{Z}$-graded rings.

On the ring theoretical level the graded theory offers a new interplay between graded and ungraded (relative) invariants. Chapter VI is mainly concerned with the study of the graded Brauer group of a graded ring and with the cohomological interpretation of this group. The relation between the graded Brauer group $Br^g(R)$ of a graded ring R and the Brauer group Br(R) depends on the ungraded structure of R but also very heavily on the gradation of R. In Section VI.5. we study certain classes of graded rings like gr-Dedekind rings or gr-local rings for which Br^g is well behaved. It is worthwhile pointing out that, by changing the gradation on R, one may change $Br^g(R)$, and sometimes the graded techniques allow to describe the full Brauer group of R as the union of graded Brauer groups of R with suitably changed gradations. A similar interplay exists between graded cohomology and usual (Amitsur) cohomology (Section VI.6.).

The final chapter is devoted to some geometrical applications. Blending our theory of (relative) graded Brauer groups with some of A. Grothendieck's results on Brauer groups of schemes and varieties we obtain a ring theoretical description of the geometrically defined Brauer groups of projective varieties. We pay special attention to the case of projective curves. The Mayer-Vietoris sequences for graded Brauer groups we derived in Section VI.4. allow us to deal also with projective curves having singular points.

F. Van Oystaeyen
A. Verschoren

ACKNOWLEDGMENTS

The National Foundation for Scientific Research (N.F.W.O.) and the University of Antwerp, U.I.A., provided both financial support and job.

The colleagues of the department of Mathematics at U.I.A. have been maintaining all these years an unlikely stable atmosphere of supporting friendship.

L. Le Bruyn and S. Casenepeel contributed some results and we benefitted from their comments and the discussions we had.

R. Hoobler has put us back on the right track, when we occasionally got lost in some geometrical web of applications.

We thank J. Murre for stimulating us to further the study of Brauer groups of projective varieties and also for the hospitality at the University of Leiden he extended to us on several occasions.

Marc Nordin and Jan Van Geel contributed directly by taking certain burdens from our shoulders, thus enabling us to get some work done.

Support through thick and thin, a pun if the authors' pictures had been included, came from our wives, family and friends.

The type-work in this is incredibly good and we are really very glad that Nina helped us out with this once again.

And Eveline never cried while papa worked, thus providing him with some necessary breaks.

CONTENTS

RELATIVE INVARIANTS OF RINGS

CHAPTER I
GENERALITIES

I.1. Localization.

Localization, together with its dual "globalization", is a well-known technique in the theory of rings and modules and it plays a fundamental part in many problems in algebraic geometry. This short section presents a crisp survey of the more formal aspects of localization in the abstract setting of torsion theories. For full detail on torsion theories the reader may consult P. Gabriel and O. Goldman's "classical" papers [40] and [44], or the monographs by J. Golan [43], B. Stenström [81], F. Van Oystaeyen [88], or F. Van Oystaeyen and A. Verschoren [94].

Let R be an associative ring with unit and let R-mod be the category of left (unitary) R-modules. A torsion theory for R, or on R-mod, is a pair of classes of left R-modules $(\mathcal{T},\mathcal{F})$ such that :

T.T.1. : For all $T \in \mathcal{T}$, $F \in \mathcal{F}$, $\mathrm{Hom}_R(T,F) = 0$.

T.T.2. : If for some $M \in$ R-mod, $\mathrm{Hom}_R(M,F) = 0$ holds for all $F \in \mathcal{F}$ then $M \in \mathcal{T}$.

T.T.3. : If for some $M \in$ R-mod, $\mathrm{Hom}_R(T,M) = 0$ holds for all $T \in \mathcal{T}$ then $M \in \mathcal{F}$.

The elements of $\mathcal{T}$ are called torsion modules while the elements of $\mathcal{F}$ are said to be torsion free modules. A torsion theory $(\mathcal{T},\mathcal{F})$ is said to be <u>hereditary</u> if and only if T is closed under taking subobjects. If (T,F) is a hereditary torsion theory then T is closed under taking coproducts, quotient objects and subobjects, extensions, whilst F is closed under taking products, subobjects, isomorphisms and injective hulls. Hereditary torsion theories correspond bijectively to <u>idempotent kernel functors</u> in R-mod, i.e. left exact subfunctors κ of the identity on R-mod such that $\kappa(M/\kappa(M)) = 0$ for all $M \in$ R-mod. The torsion theory corresponding to κ, (T_κ, F_κ) say, is given by taking for T_κ the class of left R-modules M such that $\kappa(M) = M$ whilst F_κ consists of those left R-modules N such that $\kappa(N) = 0$. An $M \in T_\kappa$ is said to be κ-torsion whereas an $M \in F_\kappa$ is said to be κ-torsion free.

Unless otherwise mentioned all torsion theories considered here are hereditary and all kernel functors are idempotent, so we omit these prefixes in the sequel. If κ is a kernel functor in R-mod, then $E \in$ R-mod is said to be κ-injective if every exact diagram in R-mod :

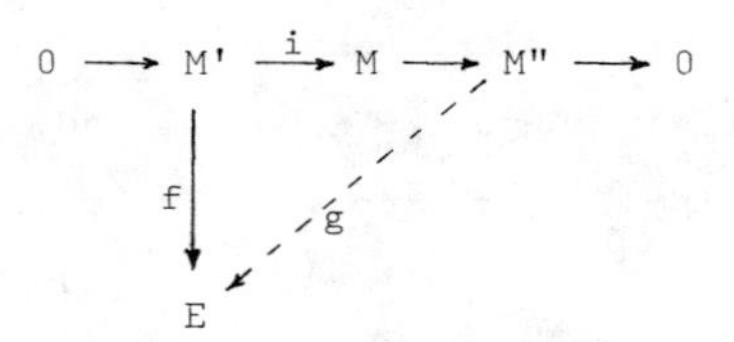

with $M'' \in T_\kappa$, may be completed by an R-linear $g : M \to E$ such that $gi = f$. We say that E is <u>κ-closed</u>, or <u>faithfully κ-injective</u>, if g as above is unique as such. Clearly, E is κ-closed if and only if E is κ-injective and κ-torsion free. The class of κ-closed left R-modules forms a full subcategory of R-mod, which we shall denote by (R,κ)-mod. This subcategory is called the <u>quotient category</u> of R-mod with respect to κ. Let $i_\kappa : (R,\kappa)\text{-mod} \to \text{R-mod}$ be the canonical inclusion.
To $M \in F_\kappa$ we associate a <u>κ-injective hull</u> of M which is κ-injective and such that $E/M \in T_\kappa$. It is obvious that each such κ-injective hull is in (R,κ)-mod.

I.1.1. <u>PROPOSITION</u>. *Every* $M \in F_\kappa$ *has an essentially unique* κ*-injective hull.*

PROOF. (Sketch). Let us indicate the construction of the κ-injective hull of M. Let I be an injective hull of M in R-mod, then $I \in F_\kappa$. Consider the exact sequence :

$$0 \longrightarrow M \longrightarrow I \xrightarrow[\pi]{} I/M \longrightarrow 0.$$

Define $E = \pi^{-1}(\kappa(I/M))$; clearly $E \in F_\kappa$. From the isomorphism $I/E \cong (I/M)/\kappa(I/M)$ it follows that $\kappa(I/E) = 0$. Since I is injective, hence κ-injective, and $I/E \in F_\kappa$, it follows that E is κ-injective. Now it follows from $E/M \cong \kappa(I/M)$ that E is a κ-injective hull of M and one easily verifies that it is unique up to isomorphism. □

The κ-injective hull of $M \in F_\kappa$ will be denoted by $E_\kappa(M)$.

I.1.2. THEOREM. *The canonical inclusion* i_κ *has a left adjoint* $\underline{\underline{a}}_\kappa$.

PROOF. If $M \in$ R-mod, put $\underline{\underline{a}}_\kappa(M) = E_\kappa(M/\kappa(M))$. This defines a functor $\underline{\underline{a}}_\kappa$: R-mod $\to$ (R,κ)-mod. Indeed, if $h : M \to N$ is a morphism in R-mod then $\underline{\underline{a}}_\kappa(h) : \underline{\underline{a}}_\kappa(M) \to \underline{\underline{a}}_\kappa(N)$ is the unique extension to $\underline{\underline{a}}_\kappa(M)$ of the induced morphism $h' : M/\kappa(M) \to N/\kappa(N)$. Now, if $f : M \to i_\kappa(N)$ is an arbitrary R-linear map, where $M \in$ R-mod, $N \in (R,\kappa)$-mod then f induces an R-linear map $f' : M/\kappa(M) \to i_\kappa(N)$. The fact that $\underline{\underline{a}}_\kappa i_\kappa(N)$ is κ-closed, while $\underline{\underline{a}}_\kappa(M)/(M/\kappa(M))$ is κ-torsion, implies that f' extends to $f_1 : \underline{\underline{a}}_\kappa(M) \to \underline{\underline{a}}_\kappa i_\kappa(N) = N$. Finally, it is easily verified that we obtain an isomorphism $\mathrm{Hom}_R(M, i_\kappa(N)) \cong \mathrm{Hom}_{(R,\kappa)\text{-mod}}(\underline{\underline{a}}_\kappa(M), N)$ and this isomorphism is actually a functorial isomorphism. □

We write $Q_\kappa = i_\kappa \underline{\underline{a}}_\kappa$. Exactness of the functor $\underline{\underline{a}}_\kappa$ entails that Q_κ is a left exact endofunctor of R-mod, which is not necessarily exact in general. Given $M \in$ R-mod, then $Q_\kappa(M) \in$ R-mod together with the canonical morphism $j_\kappa : M \to Q_\kappa(M)$, is called the <u>module of quotients of M at κ</u>.
If $M = {}_R R$ then it is possible to show that $Q_\kappa(R)$ is endowed with a natural unique ringstructure which makes j_κ into a ring morphism. If $M \in$ R-mod, then $Q_\kappa(M) \in Q_\kappa(R)$-mod in a canonical way.

To a kernel functor κ in R-mod we associate $\mathcal{L}(\kappa)$, the class of left ideals L of R such that $\kappa(R/L) = R/L$. One easily checks the following properties :

I.F.1. : If $L \in \mathcal{L}(\kappa)$ and $L \subset L'$ then $L' \in \mathcal{L}(\kappa)$.

I.F.2. : If L and L' are in $\mathcal{L}(\kappa)$ then $L \cap L' \in \mathcal{L}(\kappa)$.

I.F.3. : If $L \in \mathcal{L}(\kappa)$ and $x \in R$, then $(I : x) \in \mathcal{L}(\kappa)$.

I.F.4. : If L is a left ideal of R and $I \in \mathcal{L}(\kappa)$ is such that $(L : x) \in \mathcal{L}(\kappa)$ for all $x \in I$, then $L \in \mathcal{L}(\kappa)$.

Given a set $\mathcal{L}$ of left ideals of R, satisfying the properties I.F.1.,2., 3.,4., then we determine a kernel functor κ as follows. To an $M \in$ R-mod we associate $\kappa(M) = \{m \in M; \mathrm{Ann}_R(m) \in \mathcal{L}\}$, and we may verify that $\mathcal{L}(\kappa) = \mathcal{L}$. The set of left ideals $\mathcal{L}(\kappa)$ is called the <u>idempotent filter</u> of κ (or the <u>Gabriel topology</u> of κ). The filter associated to κ may be used to determine the modules of quotients at κ as follows :

$$Q_\kappa(M) \cong \varinjlim_{L \in \mathcal{L}(\kappa)} \mathrm{Hom}_R(L, M/\kappa(M)),$$

cf. [44].

It is clear that exactness of Q_κ, whenever it occurs, will be a very important and useful property of the localization.

I.1.3. <u>THEOREM</u>. *For any kernel functor* κ *on* R-*mod, the following statements are equivalent :*

(T)1. Q_κ *is exact and commutes with direct sums.*

(T)2. *Every* $M \in Q_\kappa(R)$-*mod is* κ-*torsion free (considering* $Q_\kappa(R)$-*modules as* R-*modules via the ring morphism* $j_\kappa : R \to Q_\kappa(R)$).

(T)3. $Q_\kappa(R)$ *is flat as a right* R-*module and* j_κ *is an epimorphism in the category of rings.*

(T)4. *Every* $Q_\kappa(R)$-*module is* κ-*closed, i.e. the full subcategories* (R,κ)-*mod and* $Q_\kappa(R)$-*mod of* R-*mod be identified in* R-*mod.*

(T)5. *For every* $L \in \mathcal{L}(\kappa)$ *we have that* $Q_\kappa(R) j_\kappa(L) = Q_\kappa(R)$.

<u>PROOF</u>. cf. [43], [44] or [94]. □

If κ has one of the equivalent properties mentioned in Theorem I.1.3. then we say that the torsion theory is <u>perfect</u>, or that κ has <u>property (T)</u>, or that κ is a <u>T-functor</u>. If κ has property (T) then κ has finite type, in the sense that every $L \in L(\kappa)$ contains an $L' \in L(\kappa)$ such that L' is a finitely generated left ideal. Even if R is commutative, there do exist examples of kernel functors κ which do not have property (T), cf. O. Goldman [44]. However, in many cases there exist "enough" T-functors on R-mod; this can be made precise by introducing certain operations in the *set* (behold !) of kernel functors for R. If σ,κ are kernel functors in R-mod then we say that $\sigma \leq \kappa$ if $\sigma(M) \subset \kappa(M)$ for all $M \in$ R-mod, or equivalently if $T_\sigma \subset T_\kappa$, or equivalently if $L(\sigma) \subset L(\kappa)$. If $\{\kappa_\alpha;\alpha \in A\}$ is a family of kernel functors in R-mod then we define

$$\kappa = \inf_{\alpha\in A} \{\kappa_\alpha\} = \bigwedge_{\alpha\in A} \kappa_\alpha$$

by $L(\kappa) = \bigcap_{\alpha\in A} L(\kappa_\alpha)$. On the other hand $\sigma = \sup_{\alpha\in A}\{\kappa_\alpha\} = \bigvee_{\alpha\in A} \kappa_\alpha$ is defined by taking for $L(\sigma)$ the idempotent filter generated by $\bigcup_{\alpha\in A} L(\kappa_\alpha)$. It is not hard to verify that κ,σ thus defined are indeed idempotent kernel functors in R-mod. These operations make the set of kernel functors on R-mod, denoted by $K(R)$, into a complete distributive (Brouwerian) lattice.

Many examples of kernel functors have been used in ring theory but also in category theory, etc... Let us point out: the localization at a prime ideal of a left Noetherian ring used by J. Lambek and G. Michler, the so-called Goldie torsion theory corresponding to the singular radical, symmetric kernel functors as introduced by D. Murdoch and F. Van Oystaeyen in [63], and finally the sheafification functor $\underline{\underline{a}}$ (which is a localization functor in the category of presheaves of modules over a [(pre-) sheaf of rings] and similar examples studied by F. Van Oystaeyen, A. Verschoren in [94]. Two special examples will turn up frequently throughout this book, so we will describe them in more detail here.

Let I be a left finitely generated ideal (ideal means two-sided ideal !) of R and let $L(I)$ be the filter generated by $\{I^n;n \in \mathbb{N}\}$. One easily verifies that $L(I)$ is an idempotent filter; the associated kernel functor will be denoted by κ_I. So if R is a left Noetherian ring then every ideal I corresponds to an idempotent kernel functor κ_I. The localization functor at κ_I will be denoted by Q_I.

Let P be a prime ideal of R and let $L(R-P)$ be the filter of left ideals of R generated by the ideals of R which are not contained in P. If R is left Noetherian or commutative then $L(R-P)$ is an idempotent filter, and the kernel functor corresponding to $L(R-P)$ will be denoted by κ_{R-P}, or sometimes just κ_P. The localization functor corresponding to κ_{R-P} is denoted by Q_{R-P}, or Q_P.

Neither κ_I, nor κ_{R-P} needs to be a (T)-functor. However, if R is commutative then Q_{R-P} is just the usual localization at P and then κ_P has property (T), for all prime ideals P of R. If I is a principal ideal, generated by $f \in R$ say, then Q_I is just localization of R at the multiplicative system $\{1,f,f^2,\ldots\}$. In the commutative case, but also in the noncommutative case, cf. [95], the kernel functors κ_I, κ_{R-P} play a fundamental role in algebraic geometry. We will come back to this later.

If R is a commutative ring and κ is a kernel functor in R-mod, then we write $C(\kappa)$ for the set of ideals of R maximal with respect to the property of not being contained in $L(\kappa)$. A well-known standard trick allows to deduce that $C(\kappa)$ consists of prime ideals of R.

I.1.4. PROPOSITION. *Let* R *be a commutative ring and let* κ *be an arbitrary kernel functor in* R-*mod then*

$$\kappa = \inf\{\kappa_{R-P}; P \in C(\kappa)\}.$$

PROOF. Cf. [63] or [88] for the proof of a more general statement concerning symmetric kernel functors. □

Other properties of rings and modules of quotients at kernel functors that we might need in the sequel will be introduced or recalled "on the spot".

I.2. The Picard group of a Commutative Ring.

In this section we briefly survey some properties of invertible and divisorial ideals, and recall some basic facts about reflexive and projective modules. We omit some proofs, not only because they are just old hat, but mainly because these results may be viewed as special cases of the general theory in Chapter II. On the other hand we hope that this survey makes Chapter II more intelligible to those who do not feel the omitted proofs to be old hat to them. Main references for this section are H. Bass [12], N. Bourbaki [15], and R. Fossum [36].

Let C be a commutative ring. A C-module P is projective and finitely generated if and only if P is finitely presented and flat, if and only if P is finitely presented and locally free. For any $M \in C$-mod we define the <u>dual of M</u> to be $M^* = \mathrm{Hom}_C(M,C)$. If P and M are as above then we have a natural transformation

$$d_P : P^* \underset{C}{\otimes} M \to \mathrm{Hom}_C(P,M) : d_P(q\otimes m)(p) = mq(p)$$

for $q \in P^*$, $m \in M$, $p \in P$. Since a finitely generated projective C-module is locally free the (local) rank function will be a handy tool in studying these modules.
If $C \to D$ is a homomorphism of commutative rings and $P \in C$-mod is finitely generated and projective then the natural homomorphism :

$$\mathrm{Hom}_C(P,M) \underset{C}{\otimes} D \to \mathrm{Hom}_D(P \underset{C}{\otimes} D, M \underset{C}{\otimes} D)$$

is an isomorphism, for every $M \in C$-mod. In particular $P^* \underset{C}{\otimes} D \cong (P \underset{C}{\otimes} D)^*$.

I.2.1. PROPOSITION. *If* $M,N \in C$-*mod are such that* $M \otimes_C N \cong C^n$ *for some positive integer* n, *then* M *and* N *are finitely generated faithfully projective* C-*modules.*

PROOF. Cf. [12]. □

A C-module M is said to be invertible if there exists an $N \in C$-mod such that $M \otimes_C N \cong C$. It follows that an invertible module M is necessarily finitely generated and faithfully projective; moreover one may easily verify that $N \cong M^*$ follows. Using the rank function it is easy to derive :

I.2.2. PROPOSITION. *The following conditions on* $M \in C$-*mod are equivalent :*

1. M *is invertible.*
2. M *is finitely generated projective and has constant rank one.*
3. M *is finitely generated and* $Q_p(M) \cong Q_p(C)$ *for all maximal ideals* p *of* C. □

On the isomorphism classes of invertible C-modules we have an operation $[P].[Q] = [P \otimes_C Q]$, (where $[-]$ denotes the isomorphism class), which defines a group law on the set of isomorphism classes of invertible C-modules. This group is called the Picard group of C and it is denoted by Pic(C). If $f : C \to D$ is a homomorphism of commutative rings then $\mathrm{Pic}(f) : \mathrm{Pic}(C) \to \mathrm{Pic}(D)$ sending $[P]$ to $[P \otimes_C D]$ is a group homomorphism and so Pic(-) is a functor.

Let κ be a T-functor on C-mod such that $\kappa(C) = 0$ and consider $j_\kappa : C \to Q_\kappa(C)$, which is monomorphic in this case. Any C-submodule $M \subset Q_\kappa(C)$ such that $M \cap C \in L(\kappa)$ is said to be a κ-open submodule. Since $L(\kappa)$ is an idempotent filter one easily checks the following properties : if M and N are κ-open then $M \cap N$, $M + N$ and $M.N$ are κ-open.
For example if κ is just localization at a multiplicatively closed set

of nonzero divisors S in C then κ-open means just "nondegenerate" in Proposition 7.7 of [12] may be generalized to:

I.2.3. PROPOSITION. *Let κ be a T-functor in C-mod such that $\kappa(C) = 0$ and let M,N be κ-open submodules of $Q_\kappa(C)$. Write*

$$(N : M) = \{x \in Q_\kappa(C); xM \subset N\}.$$

The natural homomorphism $(N : M) \to \mathrm{Hom}_C(M,N)$ is an isomorphism. Consequently, $M^{**} \cong (C : (C : M))$ and M^* is reflexive, i.e. $(M^*)^{**} \cong M^*$.

PROOF. The homomorphism $(N : M) \to \mathrm{Hom}_C(M,N)$ is given by $x \mapsto m_x(a) = ax$ for all $a \in M$. If $m_x(M) = 0$ then $(M \cap C)x = 0$ yields $x \in \kappa(Q_\kappa(C)) = 0$; thus $(N : M) \to \mathrm{Hom}_C(M,N)$ is monomorphic. To establish surjectivity of this map, consider $f \in \mathrm{Hom}_C(M,N)$. By localization we find that $Q_\kappa(f) : Q_\kappa(C) \to Q_\kappa(C)$, is given by multiplication by some $y \in Q_\kappa(C)$, i.e. $Q_\kappa(f)c = cy$ for all $c \in Q_\kappa(C)$. Since $Q_\kappa(f)|M = f$ it follows that $Q_\kappa(f)$ maps M to N, i.e. $cM \subset N$ or $c \in (N : M)$ follows. The last assertions in the proposition are easily verified. □

A C-submodule M of $Q_\kappa(C)$ is said to be an invertible submodule of $Q_\kappa(C)$ if there exists a C-submodule N of $Q_\kappa(C)$ such that $M.N = Q_\kappa(C)$. By property T for κ we have that $M = Q_\kappa(M \cap C) = Q_\kappa(C)\ j_\kappa\ (M \cap C)$. Thus $M.N = Q_\kappa(C)$ entails that $Q_\kappa(C)\ j_\kappa\ ((M \cap C)(N \cap C)) = M.N = Q_\kappa(C)$ and therefore $(M \cap C)(N \cap C) \in L(\kappa)$, meaning that M and N are both κ-open.
If χ is the maximal T-functor (note : sup of T-functors is a T-functor, cf. [88] for a more general non-commutative statement) for which C is χ-torsion free, then an ideal I of C is said to be invertible if it is an invertible submodule of $Q_\chi(C)$.

I.2.4. THEOREM. *Let κ be a T-functor such that C is κ-torsion free and let M be a κ-open C-submodule of $Q_\kappa(C)$. The following conditions are equivalent :*

1. *M is an invertible submodule of $Q_\kappa(C)$*

2. M *is finitely generated and projective*

3. $M \in \mathrm{Pic}(C)$

4. M *is finitely generated and* $Q_p(M)$ *is cyclic over* $Q_p(C)$ *for every maximal ideal* p *of* C.

PROOF. 1. ⇒ 2. Standard argument using a decomposition of $1 \in M.N = C$.
2.⇒3. Since κ is a T-functor it has finite type; in particular, from $M \cap C \in L(\kappa)$ it follows that there is an $I_1 \in L(\kappa)$ with $I_1 \subset M \cap C$ and I_1 is finitely generated. Since M is finitely generated, there is an $I_2 \in L(\kappa)$ such that $I_2 M \subset j_\kappa(C)$. Take $I \in L(\kappa)$, finitely generated, and such that $I \subset I_1 \cap I_2$. Given a prime ideal p of C, there is a prime ideal q of C such that $q \subset p$ but $I \not\subset q$. Indeed, if not then $j_p(I)$ would be in the nilradical of $Q_p(C)$, hence $I^n \subset \kappa_p(C)$ for some $n \in \mathbb{N}$. The latter yields $JI^n = 0$ for some $J \not\subset p$ because I^n is finitely generated. Now $I^n \in L(\kappa)$ and $JI^n = 0$ contradicts the fact that $\kappa(C) = 0$. Note that $\kappa_I \leqslant \kappa$ and $\kappa_I \leqslant \kappa_q$. If we localize at κ_I we have $Q_I(M) = Q_I(C)$ because $IM \subset C$ and $IC \subset M$. From $\kappa_I \leqslant \kappa_q$ it follows that Q_q is obtained by localizing first at κ_I and then at the kernel functor induced by κ_q in $Q_I(C)$-mod. Consequently $Q_q(M) \cong Q_q(C)$ or $Q_q(M)$ is free of rank one over $Q_q(C)$. From $q \subset p$ it follows that $Q_q(M) = Q_{q_p}(Q_p(M))$ where q_p is the prime ideal of $Q_p(C)$ corresponding to q. So finally $Q_p(M)$ has to be free of rank one over $Q_p(C)$.
3. ⇒ 2. and 3. ⇒ 4. are obvious.
4. ⇒ 3. Since M is κ-open it is a faithful C-module. Indeed, if $Mc = 0$ then $(M \cap C)c = 0$ yields $c \in \kappa(C) = 0$. As M is finitely generated and faithful, the same properties hold for $Q_p(M)$ as one easily checks. But then the hypothesis that $Q_p(M)$ is a cyclic $Q_p(C)$-module entails that $Q_p(M) \cong Q_p(C)$.
2. ⇒ 1. By Proposition I.2.3. we may identify M^* with $(C : M)$, so we find $m_i \in M$, $n_i \in (C : M)$ such that $m = \sum_i m_i n_i m$ (i varies in some finite set). Since M is faithful it follows that $\sum_i m_i n_i = 1$ and then $M(C : M) = C$. □
The proof given is a slight modification of the proof of Theorem 7.8.

in [ ], i.e. in particular the verification of 2. ⇒ 3. had to be modified.

I.2.5. PROPOSITION. *Let* M,N *be* C-*submodules of* $Q_\kappa(C)$ *with* M *invertible. Then* $N = (N : M)M$, $(N : M) = N(C : M)$ *and* $MN \cong M \otimes_C N$.

PROOF. Cf. Proposition 7.9. in [12] utilizing exactness of Q_κ and $Q_\kappa(T) = Q_\kappa(C) \otimes_C T$ for every $T \in C$-mod. □

We denote by Inv(C,κ) the set of invertible C-submodules of $Q_\kappa(C)$. It is clear that Inv(C,κ) is a group under multiplication and we have a group homomorphism, Inv(C,κ) → Pic(C) given by M ↦ [M].

If one observes that a finitely generated projective C-module P is κ-torsion free whenever C is κ-torsion free, (write P as a direct summand of C^m for some $m \in \mathbb{N}$) hence $P \to Q_\kappa(P) = Q_\kappa(C) \otimes_C P$ is injective as well, it is not hard to verify (along the lines of Proposition 7.10 in [12]) that we have an exact sequence :

$$0 \to U(C) \to U(Q_\kappa(C)) \to \mathrm{Inv}(C,\kappa) \to \mathrm{Pic}(C) \to \mathrm{Pic}(Q_\kappa(C)).$$

I.2.6. Remark. Not every T-functor κ as above may be obtained by localizing at a multiplicatively closed set of the ring C. Indeed, even if C is a Dedekind domain (i.e. every κ will be a T-functor) one may consider the filter $\mathcal{L}(I)$ where I is an ideal representing a non-torsion element of Pic(C) (such examples exist) and it is straightforward to verify that κ_I is not obtained as a localization at a multiplicatively closed set. The slight generalization we introduced allows to apply these techniques to arbitrary affine open sets of Spec(C), not necessarily to basic open sets !

For constructions and properties of reflexive modules over Noetherian domains we refer to N. Bourbaki [15]. Recall that a ring R is said to have global dimension less than n if every R-module M admits a projective resolution of length less than n. A commutative local ring is said to be regular if it is Noetherian and has finite global dimension. If R is a commutative Noetherian ring then gl.dim R = $\sup_p\{\text{gl.dim } Q_p(R)\}$, p ranging over the maximal (or over the prime) ideals of R. We say that R is regular if every localization $Q_p(R)$ at a maximal ideal p of R is a regular local ring. It is well-known that a regular local ring is factorial.

I.2.7. PROPOSITION. *Let C be a commutative Noetherian domain and let D be a C-algebra which is flat as a C-module. Let M,N ∈ C-mod be finitely generated, then :*

1. $\text{Hom}_D(D \underset{C}{\otimes} M, D \underset{C}{\otimes} N) \cong D \underset{C}{\otimes} \text{Hom}_C(M,N)$
2. *If M is reflexive then* $D \underset{C}{\otimes} M$ *is reflexive*
3. *If* $Q_p(M)$ *is a projective* $Q_p(C)$*-module for all prime ideals p of C then M is projective.* □

I.2.8. PROPOSITION. *If M is a finitely generated reflexive module over a regular domain C such that* $\text{End}_C(M)$ *is a projective C-module, then M is projective.* □

I.2.9. THEOREM. *Let C be a Noetherian domain of gl. dimension at most two then each finitely generated reflexive C-module is projective.* □

Some of the results mentioned above will be used in connection with the theory of (maximal) orders over integrally closed domains. We will come back to these applications later.

I.3. The Brauer Group of a Commutative Ring.

In this section we recall some definitions and basic properties of Azumaya algebras and the Brauer group. For adequate references we refer to the checklist on Brauer groups compiled by A. Verschoren in [101]. For an introduction to the theory one may look up : F. De Meyer and

E. Ingraham [33], M. Knus and M. Ojanguren [56], M. Orzech and C. Small [68].

Throughout this section C is a commutative ring with unit and by a C-algebra A we will mean a ring A containing C in its center.

Let A be a C-algebra and let A^0 be the opposite algebra of A, i.e; the C-module underlying A but equiped with the multiplication $a^0.b^0 = (ba)^0$, then $A^e = A \otimes_C A^0$ is called the <u>enveloping algebra</u> of A. There is a canonical epimorphism of left A^e-modules $m_A : A^e \to A$, $m_A(a \otimes_C b^0) = ab$. We say that A is C-separable if m_A splits in A^e-mod, i.e. if A is a projective left A^e-module. Let $j : A \to A^e$ be an A^e-linear map splitting m_A. It is easily verified that A is C-separable if and only if there exists an $e \in A^e$ such that $m_A(e) = 1$ and $(1 \otimes a^0)e = (a \otimes 1^0)e$ for all $a \in A$. Such element e is then necessarily idempotent, and it is called a <u>separability idempotent</u> for A. For every left A^e-module N we define $N^{(A)}$ to be the C-module $\{x \in N; xa = ax \text{ for all } a \in A\}$. It is clear that $N^{(A)} = eN$, where e is the separability idempotent for A. A consequence of this is that an epimorphic image B of the C-separable algebra A is again separable over the image of C and moreover the center of B is the image of Z(A), the center of A.

I.3.1. <u>LEMMA</u>. *Let* A *be a* C*-separable algebra. Consider a left* A*-module* M *which is a projective* C*-module by restriction of scalars, then* M *is a projective left* A*-module.* □

The following result stems from O. Villamayor, D. Zelinski [102] :

I.3.2. <u>PROPOSITION</u>. *A* C*-separable algebra* A *which is projective in* C*-mod is a finitely generated* C*-module.* □

If C = Z(A) and A is C-separable then we say that A is an <u>Azumaya algebra</u> (over C), or that A is central separable (these algebras were called maximally central algebras by G. Azumaya [11]). In this case the assumptions in Proposition I.3.2. are always fulfilled, indeed we have :

I.3.3. THEOREM. *If* A *is an Azumaya algebra over* C *then* A *is a finitely generated projective* C-*module and* C *is a direct summand of* A. □

The consequences of this theorem make Azumaya algebras very easy to handle in most senses. If I is an ideal in the Azumaya algebra A over C then $I = A(I \cap C)$ and conversely if I' is an ideal of C then $AI' \cap C = I'$. In other words, there is a bijective correspondence between ideals of C and ideals of A. More can be said. The categories A^e-mod and C-mod are equivalent, equivalences being given by : $(\ )^{(A)} : A^e\text{-mod} \to C\text{-mod}$, $M \mapsto M^{(A)}$, and $(\ ) \underset{C}{\otimes} A : C\text{-mod} \to A^e\text{-mod}$, $N \mapsto N \underset{C}{\otimes} A$. Note that A^e-mod may be identified with the category of two-sided A-modules. So for an Azumaya algebra A every two-sided A-module M is an Artin-bimodule in the sense of [7] i.e. M is generated as a left (right) A-module by $M^{(A)}$.

The following theorem provides a useful characterization of Azumaya algebras :

I.3.4. THEOREM. *For a* C-*algebra* A *the following conditions are equivalent :*

1. A *is an Azumaya algebra over* C;
2. A *is a faithfully projective* C-*module and the natural* C-*algebra morphism* $\eta_A : A^e \to \mathrm{End}_C(A)$ *is an isomorphism.* □

For an Azumaya algebra A over C, every kernel functor κ is symmetric, i.e. $L(\kappa)$ has a filterbasis consisting of ideals. Utilizing the properties of the functor $(\ )^{(A)}$ one easily deduces that κ is central i.e. the kernel functor κ_C induced by κ in C-mod ($L(\kappa_C)$ = {I ideal of C such that $AI \in L(\kappa)$}) is such that the localization functor Q_κ coincides with the functor induced by Q_{κ_C} in A-mod. Moreover κ is a T-functor if and only if κ_C is a T-functor.

Since an Azumaya algebra is necessarily a finite module over its center it follows that A is a P.I. algebra in the sense of [71], or

[7], ... The Artin-Procesi theorem provides an intrinsic characterization of Azumaya algebras in the class of P.I. rings :

I.3.5. THEOREM. *Let* R *be a semiprime P.I. ring with center* C, *then* R *is an Azumaya algebra (of constant rank) if and only if for every maximal ideal* P *of* R, *p.i. deg* P = *p.i. deg* R.

PROOF. Cf. [7] or [71]. □

We do not need to go into the theory of P.I. algebras, the above theorem is just mentioned for "completeness" sake; actually the fact that R will usually be a finitely generated C-module allows to use the following criteria :

I.3.6. THEOREM. *Let* A *be* C*-algebra which is finitely generated as a* C*-module. The following conditions are equivalent :*

1. A *is* C*-separable*;
2. A/pA *is* C/pC*-separable for every maximal ideal* p *of* C;
3. $Q_p(A)$ *is* $Q_p(C)$*-separable for every maximal ideal* p *of* C;
4. $Q_p(A)$ *is* $Q_p(C)$*-separable for every prime ideal* p *of* C;
5. $Q_p(A)$ *is* $Q_p(C)$*-separable for every prime ideal* p *of* A, *where* $p = P \cap C$.

□

Let us also mention that for any Azumaya algebra A over C there exists a Noetherian subring C' of C and an Azumaya algebra A' contained in A such that $A = A' \otimes_{C'} C$. Giving a commutative ring C we can consider the set of isomorphism classes of Azumaya algebras over C, A(C) say. On A(C) we define an equivalence relation by putting A ~ B in A(C) if one of the following equivalent conditions holds :

B.E. 1 : There is a faithfully projective C-module P such that
$A \otimes_C B^0 \cong \mathrm{End}_C(P)$.

B.E. 2 : There exist faithfully projective modules P and Q such that
$A \otimes_C \mathrm{End}_C(P) \cong B \otimes_C \mathrm{End}_C(Q)$.

Note also that faithfully projective C-modules are nothing but faithful finitely generated projective C-modules i.e. progenerators of the category C-mod. Now, if A and B are Azumaya algebras over C and if D is a C-algebra then one verifies that $A \otimes_C D$ and $A \otimes_C B$ are Azumaya algebras over D, resp. C. The set $Br(C) = A(C)/\sim$ is a group with respect to the operation induced by $\otimes_C$, the neutral element in the class of C and the inverse of the class of A is the class of A^0. This group is called the Brauer group of C. From the above it follows that Br(-) is a functor : Commutative rings $\to$ Abelian groups. For more detail and the calculation of some explicit examples we refer to the extensive literature.

An extension $C \subset D$ is called a splitting ring for the Azumaya algebra A over C if there exists a faithful finitely generated projective D-module P and a D-algebra isomorphism $A \otimes_C D \cong End_D(P)$.

I.3.7. PROPOSITION. *Let A be an Azumaya algebra over C. Suppose that D is a maximal commutative subring of A and consider A as a left D-module by right multiplication. The product in A induces an isomorphism* $A \otimes_C D \cong End_D(A)$. *If A is D-projective then D is C-projective and then the rank of D over C equals the rank of A over D. If D is separable over C, then A is also D-projective.* □

Whereas over a field C = k separable maximal commutative subrings of A always exist this is certainly not true in general (for example : the quaternion algebra over $\mathbb{Z}[\sqrt{2}]$). This also shows that over a ring it will be more difficult to obtain a cohomological description of the Brauer group than over a field where the theory of crossed products, cf. [33], [68], ..., works very well. In case C is a local ring, the crossed product theory will work well because of the following theorem :

I.3.8. THEOREM. *Let C be a local ring and let A be an Azumaya algebra over C. There exists a separable maximal commutative subring D of A of the form C[x], for some* $x \in A$, *such that A is a free D-module. If* $[A : C] = n^2$, *then* $\{1, x, \ldots, x^{n-1}\}$ *is a C-basis for D.* □

In general one now proceeds by looking for not necessarily separable splitting rings which still allow the development of a cohomology theory. Here étale coverings and étale cohomology will come into the picture. We refer to [47] and also to [59] for these aspects of the theory; here it will suffice to note :

I.3.9. PROPOSITION. *The following properties are equivalent for a* C-*algebra* A :

1. A *is an Azumaya algebra over* C.
2. *For every prime ideal* p *of* C, *there exists a splitting* $Q_p(C)$-*algebra for* $Q_p(A)$, *which is separable and a free* $Q_p(C)$-*module of finite rank. Moreover,* $Q_p(A)$ *is free of finite rank over that splitting ring.*
3. *There is a faithfully flat splitting ring for* A. □

Another obstruction to a "nice" crossed product theory for Azumaya algebras resides in the fact that C-automorphisms of an Azumaya algebra A over C need not be inner in general. Actually we have an exact sequence of groups :

$$0 \to \mathrm{Inn}(A) \to \mathrm{Aut}_C(A) \xrightarrow{\alpha} \mathrm{Pic}(C)$$

where α maps an automorphism φ to the invertible C-module $({}_1A_\varphi)$ where ${}_1A_\varphi$ is the A-bimodule obtained by defining the right A-module structure on A by $x.a = x\varphi(a)$ for all $x, a \in A$. Consequently if C is semi-local then all C-automorphisms of A are inner but in general this will not be the case (cf. [68]). However, if the Azumaya algebra A over C has constant rank n^2 over C then for every C-automorphism φ of A we have that φ^n is inner. Consequently, the group $\mathrm{Out}_C(A) = \mathrm{Aut}_C(A)/\mathrm{Inn}(A)$ is an abelian n-torsion group. Note that $\mathrm{Out}_C(A)$ may be identified with the following subgroup of Pic(C) : $\{[I] \in \mathrm{Pic}(C); A \otimes_C I \cong A \text{ in } A\text{-mod}\}$. It is useful to note further that every C-automorphism φ of a maximal order Λ over C in a central simple algebra Σ over the field of fractions K of C has the property that φ^n is inner (where $n^2 = (\Sigma : K)$) in the case where C is a Krull domain.

The Brauer group of certain domains is not hard to calculate if one knows that this Brauer group embeds in the Brauer group of the field of fractions of the domain. This happens for example if C is a Dedekind ring or more surprisingly if C is a regular domain. In case the regular domain C has dimension at most two we obtain :

I.3.10. THEOREM. *If* C *is a regular domain of dimension at most two then* $Br(C) = \cap\{Br(Q_p(C));$ p *a prime ideal of* C *of height one*$\}$.

PROOF. Cf. [10] or [68]; the proof uses Theorem I.2.9. Note that the intersection of the Brauer groups is taken in Br K, K the field of fractions of C. □

If D is an extension of C then we denote by Br(D/C) the subgroup of the Brauer group consisting of the Azumaya algebras over C which may be split by D, i.e. the kernel of the map $Br(C) \to Br(D)$, $[A] \mapsto [A \otimes_C D]$. Using Amitsur cohomology one fits Br(D/C) into an exact sequence in order to obtain a cohomological description of Br(C).

I.3.11. THEOREM (*Chase-Rosenberg sequence*). *Let* D *be a faithful finitely generated projective* C*-module and a* C*-algebra, then :*

1. $H^0(D/C,U) = U(C)$, *where* U *is the "units"-functor*
2. $H^1(D/C,U) = Pic(D/C) = Ker(Pic(C) \to Pic(D))$.
3. *The following sequence is exact :*

$$1 \to H^1(D/C,U) \to Pic(C) \to H^0(D/C,Pic) \to$$
$$\to H^2(D/C,U) \xrightarrow{\theta} Br(D/C) \to H^1(D/C,Pic) \to H^3(D/C,U).$$

PROOF. Cf. [21], [33], ... □

I.3.12. COROLLARY. *If* $Pic(D) = Pic(D \otimes D) = 0$ *then* $\theta : H^2(D/C,U) \to Br(D/C)$, *is an isomorphism. Note that 1. and 2. hold if* D *is faithfully flat.* □

We do not dwell further in cohomology theory here. Let us point out that, we include a very brief survey on Galois theory of commutative rings in the beginning of Section III.3. Galois splitting rings of Azumaya algebras and related results from Galois cohomology theory will be used in Section V.4.

CHAPTER II
THE RELATIVE PICARD GROUP

Throughout this chapter, R denotes a commutative ring with unit and κ will be an idempotent kernel functor on R-mod.

II.1. LEMMA. *Given an exact diagram in* R-*mod :*

$$\begin{array}{ccccccc} 0 & \longrightarrow & K & \xrightarrow{i} & M_1 & \xrightarrow{p} & M_2 \\ & & \downarrow{\scriptstyle\gamma} & & \downarrow{\scriptstyle\alpha_1} & & \downarrow{\scriptstyle\alpha_2} \\ 0 & \longrightarrow & L & \xrightarrow[j]{} & N_1 & \xrightarrow[q]{} & N_2 \end{array}$$

such that, Ker α_1, Coker α_1, *and* Ker α_2 *are κ-torsion modules, then* Ker γ *and* Coker γ *are κ-torsion.*

PROOF. Since i injects Ker γ into Ker α_1, it follows that Ker γ is κ-torsion. Let $\tau : L \to$ Coker γ, $\pi_1 : N_1 \to$ Coker α_1 be the canonical projections. Pick $x \in$ Coker γ and write $x = \tau(l)$ for some $l \in L$. From the fact that $\pi_1 j(l)$ is contained in the κ-torsion module Coker α_1 it follows that $I\,\pi_1\,j(l) = \pi_1(j(Il)) = 0$ form some $I \in L(\kappa)$. Therefore $j(Il) \subset \text{Im}\ \alpha_1$. Let M_1' be the R-submodule of M_1 mapping to $j(Il)$. Clearly $p(M_1') \subset \text{Ker}\ \alpha_2$ and thus $p(M_1')$ is κ-torsion. Therefore $M_1'/i(K) \cap M_1'$ is κ-torsion and by taking images we get that $j(Il)/j\gamma(K) \cap j(Il)$ is κ-torsion. It follows that for every $y \in I$ we have $J\,y\,l \subset \text{Im}\ \gamma$ for some $J \in L(\kappa)$. Thus : $J\,y\,\tau(l) \subset \text{Im}\ \tau\gamma = 0$, or $y\tau(l)=yx \in \kappa(\text{Coker}\ \gamma)$. From $Ix \in \kappa(\text{Coker}\ \gamma)$, $x \in \kappa(\text{Coker}\ \gamma)$ follows. □

II.2. DEFINITION. The R-module P is said to be κ-flat if for each monomorphism $i : M' \to M$ in R-mod, the kernel of $P \otimes i : P \otimes M' \to P \otimes M$ is κ-torsion.

This will hold exactly then when the induced morphism $Q_\kappa(P \otimes i)$ is monomorphic. One easily checks that P is κ-flat if and only if for each morphism $u : M \to N$ in R-mod such that Ker u is κ-torsion, we have that Ker $P \otimes u$ is κ-torsion too. In the non-commutative case the definition of κ-flatness will be modified !

II.3. PROPOSITION. *If* P *is* κ*-flat and* E *is* κ*-injective, then* $[P,E] = \mathrm{Hom}_R(P,E)$ *is* κ*-injective.*

PROOF. Consider an exact sequence in R-mod :

$$0 \to M' \to M \to M'' \to 0,$$

where M" is κ-torsion. The κ-injectivity of (P,E) amounts to $[M,[P,E]] \to [M',[P,E]]$ being surjective. Since $[P,-]$ and $P \underset{R}{\otimes} -$ are adjoint functors we only have to check whether the mapping

$$\varphi : [P \otimes M,E] \to [P \otimes M',E],$$

is surjective. The exact sequence :

$$0 \to K \to P \otimes M' \to P \otimes M \to P \otimes M'' \to 0$$

has the property that its extremes are κ-torsion modules. Now it follows from the κ-injectivity of E that φ is surjective. □

II.4. LEMMA. P *is* κ*-flat if and only if* $Q_\kappa(P)$ *is* κ*-flat.*

PROOF. Consider a monomorphism $i : M \to N$ in R-mod and the following commutative exact diagram in R-mod (where $j_P : P \to Q_\kappa(P)$ is the canonical morphism) :

$$\begin{array}{ccccccc} 0 \to & \mathrm{Ker}\, P\otimes i & \to & P\otimes M & \longrightarrow & P\otimes N \\ & \downarrow & & \downarrow j_P\otimes M & & \downarrow j_P\otimes N \\ 0 \to & \mathrm{Ker}\, Q_\kappa(P)\otimes i & \to & Q_\kappa(P)\otimes M & \xrightarrow{Q_\kappa(P)\otimes i} & Q_\kappa(P)\otimes N \end{array}$$

In this diagram, Ker and Coker of both $j_P \otimes M$ and $j_P \otimes N$ are κ-torsion. Applying Lemma II.1. and taking into account that Ker $P \otimes i$ is also κ-torsion, the statement follows. □

II.5. LEMMA. *For* R-*modules* P *and* Q *we have :*

$$Q_\kappa(P\otimes Q) = Q_\kappa(Q_\kappa(P)\otimes Q).$$

PROOF. View Q as the cokernel of a map $R^{(I)} \to R^{(J)}$ for certain sets I,J. So we obtain an exact diagram :

$$\begin{array}{ccccccc} P\otimes R^{(I)} & \to & P\otimes R^{(J)} & \to & P\otimes Q & \to & 0 \\ \downarrow \alpha_1 & & \downarrow \alpha_2 & & \downarrow \gamma & & \\ Q_\kappa(P)\otimes R^{(I)} & \to & Q_\kappa(P)\otimes R^{(J)} & \to & Q_\kappa(P)\otimes Q & \to & 0 \end{array}$$

Here Ker and Coker of both α_1 and α_2 are κ-torsion because Ker j_P and Coker j_P are κ-torsion too.
Applying the "dual" of Lemma II.1. yields that Ker γ and Coker γ are κ-torsion and therefore γ induces an isomorphism $Q_\kappa(P\otimes Q) \cong Q_\kappa(Q_\kappa(P)\otimes Q)$. □

An R-module P is said to be <u>κ-invertible</u> if it is κ-closed and there exists a κ-closed R-module Q such that $Q_\kappa(P\otimes Q) \cong Q_\kappa(R)$. If κ is a T-functor then a κ-invertible R-module will be nothing but an invertible $Q_\kappa(R)$-module in the sense of Chapter I.

II.6. PROPOSITION. *A κ-invertible module is κ-flat.*

PROOF. Let P be κ-invertible and consider a monomorphism $i : M' \to M$. If Q is κ-invertible and such that $Q_\kappa(P \otimes Q) \to Q_\kappa(R)$ then we deduce from the exact sequence : $0 \to K \to P \otimes M' \to P \otimes M$, that the composite $Q_\kappa(Q \otimes P \otimes i)Q_\kappa(Q \otimes j) : Q_\kappa(Q \otimes K) \to Q_\kappa(Q \otimes P \otimes M)$ is the zero map. However we have that

$$Q_\kappa(Q \otimes P \otimes -) = Q_\kappa(Q_\kappa(Q \otimes P) \otimes -) = Q_\kappa(Q_\kappa(R) \otimes -) = Q_\kappa(R \otimes -) = Q_\kappa(-)$$

and the map $Q_\kappa(Q \otimes P \otimes i)$ therefore reduces to the monomorphism $Q_\kappa(i) : Q_\kappa(M') \to Q_\kappa(M)$.
Hence, $Q_\kappa(Q \otimes j)$ is the zero map and the same holds for

$$Q_\kappa(P \otimes Q_\kappa(Q \otimes j)) : Q_\kappa(P \otimes Q_\kappa(Q \otimes K)) \to Q_\kappa(P \otimes Q_\kappa(Q \otimes P \otimes M')).$$

The latter map reduces to $Q_\kappa(j) : Q_\kappa(K) \to Q_\kappa(P \otimes M')$. Because $Q_\kappa(j) = 0$, it follows that K is a κ-torsion module. □

II.7. LEMMA. *If* P *is κ-invertible, then* $\mathrm{End}_R(P) \cong Q_\kappa(R)$.

PROOF. Let the map $\theta : Q_\kappa(R) \to \mathrm{End}_R(P)$ be given by taking for $\theta(r)$ the endomorphism of P given by scalar multiplication by r. It is clear that θ is a ring morphism. If $\theta(r) = 0$ then $rP = 0$ yields $rP \otimes Q = 0$, where Q is as above. We have a commutative diagram in R-mod :

$$\begin{array}{ccc}
P \otimes Q & \xrightarrow{\varphi_r} & P \otimes Q \\
\downarrow & & \downarrow \\
Q_\kappa(P \otimes Q) & \xrightarrow{\varphi_r} & Q_\kappa(P \otimes Q) \\
\downarrow\cong & & \downarrow\cong \\
Q_\kappa(R) & \xrightarrow{\varphi_r} & Q_\kappa(R)
\end{array}$$

where φ_r is just multiplication by r. But φ_r can only be the zero map on $Q_\kappa(R)$ if $r = 0$. This shows that θ is injective. Consider $f \in \mathrm{End}_R(P)$ and the following commutative diagram in R-mod :

$$\begin{array}{ccc} Q_\kappa(Q\otimes P) & \xrightarrow{\cong} & Q_\kappa(R) \\ \downarrow{\scriptstyle Q_\kappa(Q\otimes f)} & & \downarrow{\scriptstyle g} \\ Q_\kappa(Q\otimes P) & \xrightarrow{\cong} & Q_\kappa(R) \end{array}$$

where g is the unique extension of $Q \otimes f$ to $Q_\kappa(R)$. Obviously g is just multiplication by $g(1) = r$. Define $h \in \mathrm{End}_R(P)$ by $h(p) = rp$ for all $p \in P$. Then, $Q_\kappa(Q\otimes f) = Q_\kappa(Q\otimes h) = g$ and since $\mathrm{Im}(f-h) \subset P$ it follows that $\mathrm{Im}(f-h)$ is κ-torsion free. This allows to deduce from $Q_\kappa(Q\otimes \mathrm{Im}(f-h)) = 0$ that $\mathrm{Im}(f-h) = 0$. □

II.8. <u>DEFINITION</u>. The set of isomorphism classes $[P]$ of κ-invertible R-modules P, denoted by $\mathrm{Pic}(R,\kappa)$ has a group structure with respect to the operation $[P].[P'] = [Q_\kappa(P\otimes P')]$. The unit element of this group is $[Q_\kappa(R)]$. This group is called the <u>κ-relative Picard group</u>.

II.9. <u>Examples</u>. Let us give some examples concerning κ-flatness.

1. A direct sum $\bigoplus_i P_i$ is κ-flat if and only if each P_i is κ-flat.
2. A direct limit of κ-flat modules is κ-flat.
3. A projective module is κ-flat for each κ.
4. An R-module P is said to be <u>κ-projective</u> if for each $M' \to M'' \to 0$, and $P \to M''$ there exists a submodule P' in P and a morphism $P' \to M'$ such that the following diagram commutes

   $$\begin{array}{ccccc} 0 \to & P' & \to & P & \\ & \downarrow & & \downarrow & \\ & M'' & \to & M' & \to 0 \end{array}$$

   and such that P/P' is κ-torsion.
   If P is κ-closed and κ-projective then it is also κ-flat. It is well-known that κ-projective modules (in fact ideals) play an important role in the characterization of T-functors, cf. [43], [44].
5. If R is a Krull domain then every reflexive R-module is σ-flat where $\sigma = \inf\{\kappa_{R-p}; P\in X^{(1)}(R)\}$ where $X^{(1)}(R)$ is the set of height one prime ideals of R.

In order to study the functorial behaviour of Pic(R,κ) with respect to ring homomorphisms we introduce some category theoretical machinery.

A <u>category with product</u> is a category $\underline{C}$ with a product functor, $\perp : \underline{C} \times \underline{C} \to \underline{C}$ together with commutativity and associativity isomorphisms which are coherent in the sense of MacLane [61].
A <u>product preserving functor</u> between categories with product $\underline{C}$ and $\underline{D}$ is a functor $F : \underline{C} \to \underline{D}$ such that there is a natural isomorphism $F(C_1 \perp C_2) \cong FC_1 \perp FC_2$ which is compatible with the commutativity and associativity isomorphisms. Let us recall briefly some generalities on the K-theory of these categories. Define $K_0\underline{C}$ to be the abelian group with generators [C], the isomorphism classes of objects of $\underline{C}$ and with relations $[C_1 \perp C_2] = [C_1] + [C_2]$. We also consider the category $\underline{\Omega C}$ with objects (C,α) where α is a $\underline{C}$-automorphism of C. A morphism $(C_1,\alpha) \to (C_2,\beta)$ is given by a morphism $f : C_1 \to C_1$ such that $f\alpha = \beta f$. The product $(C_1,\alpha) \perp (C_2,\beta)$ is defined to be $(C_1 \perp C_2, \alpha \perp \beta)$. We define $K_1\underline{C}$ to be $K_0\ \underline{\Omega C}$ modulo the subgroup generated by

$$[(C,\alpha\beta)] - [(C,\alpha)] - [(C,\beta)],$$

where α and β are $\underline{C}$-automorphisms of C. If $F : \underline{C} \to \underline{D}$ is a product preserving functor then $\underline{\phi F}$ denotes the category with objects (C_1,α,C_2) where C_1 and C_2 are objects of $\underline{C}$ and $\alpha : FC_1 \to FC_2$ is an isomorphism in $\underline{D}$ and with morphisms, $(f_1,f_2) : (C_1,\alpha,C_2) \to (C_1',\alpha',C_2')$, such that $f_1 : C_1 \to C_1'$, $f_2 : C_2 \to C_2'$ make the following diagram commutative :

$$\begin{array}{ccc} FC_1 & \xrightarrow{\alpha} & FC_2 \\ \downarrow Ff_1 & & \downarrow Ff_2 \\ FC_1' & \xrightarrow{\alpha'} & FC_2' \end{array}$$

We define : $(C_1,\alpha,C_2) \perp (C_1',\alpha',C_2') = (C_1 \perp C_1', \alpha \perp \alpha', C_2 \perp C_2')$ and $K_1\ \underline{\phi F}$ will be $K_0\ \underline{\phi F}$ modulo the subgroup generated by elements of the form :

$$[(C_1,\beta\alpha,C_3)] - [(C_1,\alpha,C_2)] - [(C_2,\beta,C_3)].$$

We say that a functor $F : \underline{C} \to \underline{\mathcal{D}}$ is cofinal if for a given D' in $\underline{\mathcal{D}}$ there exists a D" in $\underline{\mathcal{D}}$ together with a C in $\underline{C}$ such that $D' \perp D'' \cong FC$.
The following proposition may be found in full detail in [13].

II.10. PROPOSITION. *With notation as before :*

1. *If* $\underline{C}_0$ *is a full cofinal subcategory of* $\underline{C}$ *then the inclusion induces an isomorphism* $K_1\, \underline{C}_0 \cong K_1\, \underline{C}$.
2. *If* $F : \underline{C} \to \underline{\mathcal{D}}$ *is a cofinal product preserving functor, then there is a unique homomorphism* $K_1\, \underline{\mathcal{D}} \to K_1\, \underline{\phi F}$ *which maps the class of* (FC,α) *to the class of* (C,α,C). *The resulting sequence :*

$$K_1\, \underline{C} \to K_1\, \underline{\mathcal{D}} \to K_1\, \underline{\phi F} \to K_0\, \underline{C} \to K_0\, \underline{\mathcal{D}},$$

is exact. □

The category of κ-invertible R-modules will be denoted by $\underline{\text{Pic}}(R,\kappa)$; it is a category with product induced by $Q_\kappa(P \otimes Q) = P \perp Q$. If $f : R \to S$ is a morphism of commutative rings and κ is a kernel functor on R-mod, then κ defines a kernel functor $\overline{\kappa}$ on S-mod via the restriction of scalars functor S-mod → R-mod, by taking for the class of $\overline{\kappa}$-torsion S-modules exactly the S-modules which turn out to be κ-torsion R-modules by restriction of scalars. If $M \in$ S-mod then $Q_\kappa(M)$ is an S-module so in the commutative case we will write $Q_\kappa(M)$ for $Q_\kappa(M)$.

II.11. PROPOSITION. *Let* κ *be a kernel functor on* R-*mod and let* $f : R \to S$ *be a morphism of commutative rings. The functor*

$$Q_\kappa(- \otimes S) : \underline{\text{Pic}}(R,\kappa) \to \underline{\text{Pic}}(S,\overline{\kappa}),$$

is product preserving.

PROOF. If $P \in \underline{\text{Pic}}(R,\kappa)$ then $P \otimes -$ maps κ-torsion modules to κ-torsion modules and there is a $Q \in \underline{\text{Pic}}(R,\kappa)$ such that $Q_\kappa(P \otimes Q) \cong Q_\kappa(R)$.
Let us show that the S-modules $Q_\kappa(Q_\kappa(M) \underset{S}{\otimes} Q_\kappa(N))$ and $Q_\kappa(M \underset{S}{\otimes} N)$ are isomorphic for all $M,N \in$ S-mod. Consider the canonical map

$$f : M \otimes_S N \to Q_\kappa(M) \otimes_S Q_\kappa(N).$$

For all $P \in C(\sigma)$, f induces isomorphisms

$$f_P : Q_{R-P}(M \otimes_S N) = Q_{R-P}(M)_{Q_{R-P}(S)}\otimes Q_{R-P}(N) \to Q_{R-P}(Q_\kappa(M) \otimes_S Q_\kappa(N)),$$

and therefore f induces an isomorphism

$$Q_\kappa(f) : Q_\kappa(M \otimes_S N) \to Q_\kappa(Q_\kappa(M) \otimes_S Q_\kappa(N))$$

(see Proposition I.1.4.).
If P,Q are such that $Q_\kappa(P \otimes_R Q) \cong Q_\kappa(R)$ then the induced morphism $P \otimes_R Q \to Q_\kappa(R)$ yields a map

$$(P \otimes_R S) \otimes_S (Q \otimes_R S) = (P \otimes_R Q) \otimes_R S \to Q_\kappa(R) \otimes_R S.$$

The latter mapping gives rise to the following isomorphisms :

$$Q_\kappa(Q_\kappa(P \otimes_R S) \otimes_S Q_\kappa(Q \otimes_R S)) \cong Q_\kappa((P \otimes_R S) \otimes_S (Q \otimes_R S)) \cong Q_\kappa((P \otimes_R Q) \otimes_R S)$$
$$\cong Q_\kappa(Q_\kappa(P \otimes_R Q) \otimes_R S) \cong Q_\kappa(Q_\kappa(R) \otimes_R Q_\kappa(S)) \cong Q_\kappa(R \otimes_R Q_\kappa(S)) \cong Q_\kappa(S).$$

II.12. <u>COROLLARY</u>. *A morphism* $f : R \to S$ *of commutative rings induces a groupmorphism* $\mathrm{Pic}(f,\kappa) : \mathrm{Pic}(R,\kappa) \to \mathrm{Pic}(S,\kappa)$, *where we have denoted* $\overline{\kappa}$ *again by* κ. □

II.13. <u>LEMMA</u>. *Let* R *be a Noetherian ring and let* S *be a multiplicative subset of* R, $1 \in S$, $0 \notin S$. *Let* $M \in R$-*mod be* S-*torsion free (i.e.* $sm = 0$ *with* $s \in S$, $m \in M$ *yields* $m = 0$*). For any kernel functor* κ *on* R-*mod we have* $Q_\kappa(S^{-1}M) = S^{-1}Q_\kappa(M)$.

<u>PROOF</u>. A consequence of a general compatibility argument due to Van Oystaeyen [89]. Let us include a direct proof here.
The canonical inclusion $Q_\kappa(M) \to Q_\kappa(S^{-1}M)$, extends to an inclusion $S^{-1}Q_\kappa(M) \to S^{-1}Q_\kappa(S^{-1}M)$. We claim : $S^{-1}Q_\kappa(S^{-1}M) = Q_\kappa(S^{-1}M)$. Indeed, if $s^{-1}m \in S^{-1}Q_\kappa(S^{-1}M)$ with $s \in S$, $m \in Q_\kappa(S^{-1}M)$ then $Im \subset j_\kappa(S^{-1}M)$ (where j_κ denotes the canonical morphism $S^{-1}M \to Q_\kappa(S^{-1}M)$), for some $I \in L(\kappa)$. Thus $Im\, s^{-1} \subset S^{-1}j_\kappa(S^{-1}M)$ and $S^{-1}j_\kappa(S^{-1}M) = S^{-1}M/S^{-1}\kappa(S^{-1}M)$. It is ob-

vious that $S^{-1}\kappa(S^{-1}M) = \kappa(S^{-1}M)$ and therefore our claim does hold. So we obtain a monomorphism $S^{-1}Q_\kappa(M) \to Q_\kappa(S^{-1}M)$. Conversely, if $m \in Q_\kappa(S^{-1}M)$ then there is an $I \in L(\kappa)$ such that $Im \subset j_\kappa(S^{-1}M)$. Since I is finitely generated we may select an $s \in S$ such that $sIm \subset j_\kappa i(M)$ where $j_\kappa i(M)$ is the image of $M \to S^{-1}M \to Q_\kappa(S^{-1}M)$. Identifying M with its image in $S^{-1}M$ we get $sm \in Q_\kappa(M)$ and $m \in S^{-1}Q_\kappa(M)$. □

II.14. COROLLARY. *Let* R *be a Noetherian ring and let* S *be a multiplicative set consisting of regular elements of* R, *then the canonical map* $\mathrm{Pic}(R,\kappa) \to \mathrm{Pic}(S^{-1}R,\kappa)$ *associated to* $j_S : R \to S^{-1}R$, *maps* $[P] \in \mathrm{Pic}(R,\kappa)$ *to* $[S^{-1}P]$. □

Consider the polynomial ring $S = R[T]$ in one variable, (arguments given generalise easily to the case where T is a finite set of variables) If $P \in \underline{\mathrm{Pic}}(R,\kappa)$ then $Q_\kappa(P) \underset{R}{\otimes} S \cong Q_\kappa(P)[T]$ and the canonical map $P[T] \to Q_\kappa(P)[T]$ has κ-torsion Ker and Coker, consequently $Q_\kappa(P \underset{R}{\otimes} S) = Q_\kappa(P) \underset{R}{\otimes} S$.

II.15. LEMMA. *The functor* $\mathrm{Pic}(R,\kappa) \to \mathrm{Pic}(R[T],\kappa)$ *induced by the inclusion* $R \to R[T]$ *maps* $[P]$ *to* $[P[T]]$.

PROOF. In order to establish the lemma we have to show that P[T] is κ-injective, i.e. we have to check the existence of $\overline{f}$ making the following diagram commutative :

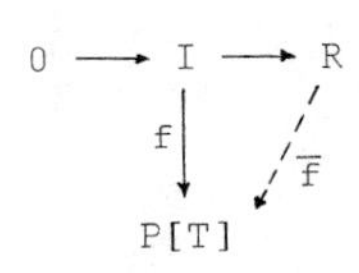

where $I \in L(\kappa)$ and $f : I \to P[T]$ is a given R-linear map. Since I is finitely generated, say $I = \sum_{i=1}^{n} R\alpha_i$, we may write $f(\alpha_i) = \sum q_{ji}T^j$. If $\sum r_i\alpha_i = \sum s_i\alpha_i \in I$ then $\sum_i r_iq_{ji} = \sum_i s_iq_{ji}$ holds for all j. For each j let us define :

$$f_j : I \to P, \qquad \sum r_i\alpha_i \mapsto \sum r_iq_{ji}.$$

Obviously each f_j is well defined and R-linear so it extends to a unique $\overline{f}_j : R \to P$, where the latter is completely determined by $\overline{f}_j(1) = \mu_j \in P$. Let $\overline{f} : R \to P[T]$ be the R-linear map determined by $\overline{f}(1) = \Sigma\, \mu_j T^j$. It is easily checked that $\overline{f}$ extends f as desired. □

II.16. <u>Note</u>. Since R[T] is a free extension of R, the fact that $Q_{\overline{\kappa}}$ and Q_κ coincide on R[T]-mod is even easier to check. Note that the above proof could be derived from elementary facts about graded localization as contained in [65].

We are now ready to apply the generalities of category theoretical nature to our situation. Let $f : R \to S$ be a morphism of commutative rings and consider the product preserving functor $F = Q_\kappa(- \underset{R}{\otimes} S)$, F : <u>Pic</u>$(R,\kappa) \to$ <u>Pic</u>(S,κ). We have seen that K_0 <u>Pic</u>$(R,\kappa) = \mathrm{Pic}(R,\kappa)$. Viewing $Q_\kappa(R)$ as a single object of a category $\langle Q_\kappa(R)\rangle$, we obtain a cofinal subcategory of <u>Pic</u>(R,κ). Indeed, for all P in <u>Pic</u>(R,κ) there is a Q in <u>Pic</u>(R,κ) such that $P \perp Q = Q_\kappa(P \otimes Q) \simeq Q_\kappa(R)$, hence K_1 <u>Pic</u>$(R,\kappa) = K_1\langle Q_\kappa(R)\rangle$. But the latter group is just the group of $\langle Q_\kappa(R)\rangle$-automorphisms of $Q_\kappa(R)$ i.e. it may be identified with $U(Q_\kappa(R))$, the units of $Q_\kappa(R)$. Writing $\mathrm{Pic}(f,\kappa)$ for K_1 <u>ϕF</u> then we obtain an exact sequence :

$$U(Q_\kappa(R)) \to U(Q_\kappa(S)) \to \mathrm{Pic}(f,\kappa) \to \mathrm{Pic}(R,\kappa) \to \mathrm{Pic}(S,\kappa).$$

Now we apply this to the following particular case. Let S be a multiplicative set in R, consisting of regular elements. We write $\mathrm{Inv}(R,S,\kappa)$ for the group of R-modules I in $S^{-1}Q_\kappa(R)$ which are κ-closed and which have the property that there exists an R-submodule J of $S^{-1}Q_\kappa(R)$ such that $Q_\kappa(IJ) = Q_\kappa(R)$. We denote by $\mathrm{Pic}(R,S,\kappa)$ the group of all (κ-closed) κ-invertible R-submodules of $S^{-1}Q_\kappa(R)$.

II.17. <u>LEMMA</u>. *If* $I \in \mathrm{Inv}(R,S,\kappa)$ *then* $I \cap S \neq \phi$.

<u>PROOF</u>. If J is as above then $S^{-1}I = S^{-1}(IS^{-1}Q_\kappa(R))$ contains $S^{-1}(IJ)$.

Hence $S^{-1}I = S^{-1}Q_\kappa(I) = Q_\kappa(S^{-1}I)$ contains $Q_\kappa(S^{-1}(IJ)) = S^{-1}Q_\sigma(R)$ and therefore $S^{-1}I = S^{-1}Q_\kappa(R)$, so consequently $I \cap S = \phi$. □

II.18. <u>LEMMA</u>. $\mathrm{Pic}(R,S,\kappa) = \mathrm{Inv}(R,S,\kappa)$.

<u>PROOF</u>. If $I \in \mathrm{Inv}(R,S,\kappa)$ then I is κ-flat. Indeed, if $Q_\kappa(IJ) = Q_\kappa(R)$ then for all $P \in C(\kappa)$ we have $Q_\kappa(IJ)_P = Q_{R-P}(R)$. So from $Q_{R-P}(IJ) = Q_{R-P}(I)Q_{R-P}(J)$ it follows that $Q_{R-P}(I)$ is invertible in $Q_{R-P}(S^{-1}R)$. Thus, $Q_{R-P}(I)$ is a projective $Q_{R-P}(R)$-module. From this κ-flatness follows by an easy local-global argument. Next, consider the canonical morphism $I \underset{R}{\otimes} J \to IJ$. By κ-flatness of I, the inclusion $J \subset Q_\kappa(S^{-1}R)$ yields a morphism
of which is κ-torsion. But the latter module may be identified with $S^{-1}I = S^{-1}Q_\kappa(R)$ because of Lemma II.17., so the image of $Q_\kappa(i)$ is exactly $Q_\kappa(IJ)$ and $Q_\kappa(i)$ is injective. It follows that $I \underset{R}{\otimes} J \to IJ$ induces an isomorphism $Q_\kappa(I \underset{R}{\otimes} J)\ Q_\kappa(IJ) = Q_\kappa(R)$. □

If S is as before, then one may prove along the lines of a similar result by H. Bass [12], that there is an exact sequence :

$$0 \to U(Q_\kappa(R)) \xrightarrow[u_1]{} U(S^{-1}Q_\kappa(R)) \xrightarrow[u_2]{} \mathrm{Pic}(R,S,\kappa) \xrightarrow[u_3]{} \mathrm{Pic}(R,\kappa) \xrightarrow[u_4]{} \mathrm{Pic}(S^{-1}R,\kappa)$$

where u_i, $i = 1,\ldots,4$, are the canonical maps. A more direct proof of the exactness of this sequence may be derived from the observation that $\mathrm{Pic}(R,S,\kappa) = \mathrm{Inv}(R,S,\kappa)$ is isomorphic to $\mathrm{Pic}(j_S,\kappa)$ where $j_S : R \to S^{-1}R$ is the canonical localization morphism, and then applying Proposition II.10.

Now consider a cartesian diagram of commutative rings :

(*)
$$\begin{array}{ccc} R & \xrightarrow{h_1} & R_1 \\ {\scriptstyle h_2}\downarrow & & \downarrow{\scriptstyle f_1} \\ R_2 & \xrightarrow{f_2} & R' \end{array}$$

i.e. $R = \{(r_1,r_2) \in R_1 \times R_2;\ f_1(r_1) = f_2(r_2)\}$. For each idempotent kernel functor κ on R-mod the above diagram gives rise to another commutative diagram :

$$(*')\qquad \begin{array}{ccc} \underline{\mathrm{Pic}}(R,\kappa) & \xrightarrow{H_1} & \underline{\mathrm{Pic}}(R_1,\kappa) \\ \downarrow H_2 & & \downarrow F_1 \\ \underline{\mathrm{Pic}}(R_2,\kappa) & \xrightarrow[F_2]{} & \underline{\mathrm{Pic}}(R',\kappa) \end{array}$$

together with a natural isomorphism $\beta : F_1H_1 \xrightarrow{\cong} F_2H_2$, where $H_i = Q_\kappa(R_i \underset{R}{\otimes} -)$, $F_i = Q_\kappa(R' \underset{R_i}{\otimes} -)$, $i = 1,2$, while β is induced by the morphisms :

$$Q_\kappa(Q_\kappa(P \underset{R}{\otimes} R_i) \underset{R_i}{\otimes} R') \cong Q_\kappa(P \underset{R}{\otimes} R').$$

Write $\underline{\mathrm{Pic}}$ for the fibre product $\underline{\mathrm{Pic}}(R_1,\kappa) \underset{\underline{\mathrm{Pic}}(R',\kappa)}{\times} \underline{\mathrm{Pic}}(R_2,\kappa)$ with objects (P_1,α,P_2), where $P_i \in \underline{\mathrm{Pic}}(R_i,\kappa)$ and α an isomorphism $F_1P_2 \xrightarrow{\cong} F_2P_2$ in $\underline{\mathrm{Pic}}(R',\kappa)$, and with morphisms $(\varphi_1,\varphi_2) : (P_1,\alpha,P_2) \to (Q_1,\beta,Q_2)$ where $\varphi_i : P_i \to Q_i$ are morphisms in $\underline{\mathrm{Pic}}(R_i,\kappa)$ making the following diagram commutative :

$$\begin{array}{ccc} F_1P_1 & \xrightarrow{\alpha} & F_2P_2 \\ \downarrow F_1\varphi_1 & & \downarrow F_2\varphi_2 \\ F_1Q_1 & \xrightarrow[\beta]{} & F_2Q_2 \end{array}$$

With these notations, we have a canonical functor :

$$T : \underline{\mathrm{Pic}}(R,\kappa) \to \underline{\mathrm{Pic}},\ P \mapsto (H_1P,\beta_P,H_2P).$$

Before stating the main theorem, we recall some definitions and elementary results from [12].

First note that, if $\varphi : R \to S$ is a morphism of commutative rings, then the induced functor $\underline{\mathrm{Pic}}(\varphi,\kappa)$ from $\underline{\mathrm{Pic}}(R,\kappa)$ to $\underline{\mathrm{Pic}}(S,\kappa)$ is a cofinal product preserving functor. This notion may be generalized as follows :

a pair of functors $F_1 : \underline{\mathcal{C}}_1 \to \underline{\mathcal{D}}$, $F_2 : \underline{\mathcal{C}}_2 \to \underline{\mathcal{D}}$ is a <u>cofinal pair</u>, if both are cofinal product preserving functors such that for every $C_2 \in \underline{\mathcal{C}}_2$ there is a C_2' in $\underline{\mathcal{C}}_2$ and C_1 in $\underline{\mathcal{C}}_1$ such that $F_2(C_2 \perp C_2') = F_1C_1$ and similarly for i replaced by $3 - i$. Again, any pair of morphisms of commutative rings $\varphi_1 : R_1 \to S$, $\varphi_2 : R_2 \to S$, determines a cofinal pair of functors $\underline{\mathrm{Pic}}(\varphi_1,\kappa)$, $\underline{\mathrm{Pic}}(\varphi_2,\kappa)$. A product preserving functor $F : \underline{\mathcal{C}} \to \underline{\mathcal{D}}$ is said to be <u>E-surjective</u> if for every C in $\underline{\mathcal{C}}$ and each δ in the commutator subgroup of $\mathrm{Aut}_{\underline{\mathcal{D}}}(FC)$ we may find a C' in $\underline{\mathcal{C}}$ and γ in the commutator subgroup of $\mathrm{Aut}_{\underline{\mathcal{C}}}(C \perp C')$ such that $F\gamma = \delta \perp 1_{FC'}$. The fact that for any P in $\underline{\mathrm{Pic}}(R,\kappa)$ we have $[P,P] \xrightarrow{\cong} Q_\kappa(R)$ entails that each morphism of commutative rings $\varphi : R \to S$ determines an E-surjective functor $\underline{\mathrm{Pic}}(\varphi,\kappa)$. Finally, a diagram of product preserving functors :

(**)

$$\begin{array}{ccc} \underline{\mathcal{B}} & \xrightarrow{H_1} & \underline{\mathcal{C}}_1 \\ {\scriptstyle H_2}\downarrow & & \downarrow{\scriptstyle F_1} \\ \underline{\mathcal{C}}_2 & \longrightarrow & \underline{\mathcal{D}} \end{array}$$

together with a natural isomorphism $\beta : F_1H_1 \xrightarrow{\cong} F_2H_2$, is said to be <u>E-surjective</u> if the following condition holds : given B in $\underline{\mathcal{B}}$ and ε in the commutator subgroup of $\mathrm{Aut}_{\underline{\mathcal{D}}}(F_1H_1B)$, there exist B' in $\underline{\mathcal{B}}$ and ε_i in the commutator subgroup of $\mathrm{Aut}_{\underline{\mathcal{C}}_i}(H_i(B \perp B'))$, $i = 1,2$, such that

$$(\varepsilon_1,\varepsilon_2) : (TB)\varepsilon \perp TB' \longrightarrow TB \perp TB'$$

is an $\underline{A}$-isomorphism where $\underline{A} = \mathcal{C}_1 \times_{\underline{\mathcal{D}}} \underline{\mathcal{C}}_2$ and $T : \underline{\mathcal{B}} \to \underline{A}$ is the canonical functor.

For categories of the type $\underline{\mathrm{Pic}}(S,\kappa)$ all ε, ε_1, ε_2 reduce to the identity and so the diagrams occurring in this case will all be E-surjective. Recall further from [12] that in case H_1 and H_2 are cofinal and (**) is an E-surjective diagram then $T : \underline{\mathcal{B}} \to \underline{A}$ is cofinal too.

II.19. <u>THEOREM</u>. *With notations as in (*) we have : the canonical functor* $T : \underline{\mathrm{Pic}}(R,\kappa) \to \underline{\mathrm{Pic}}$ *is an equivalence of categories.*

<u>PROOF</u>. The cartesian square (*) induces a square :

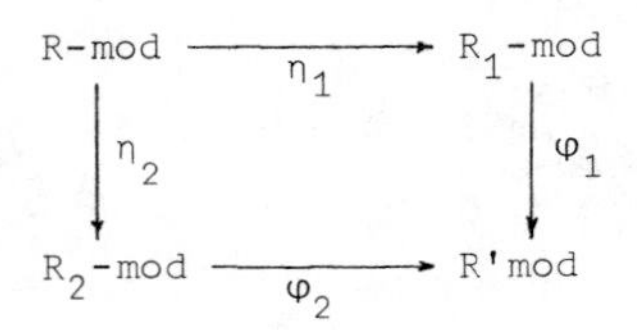

where $\eta_i = R_i \underset{R}{\otimes} -$, $\varphi_i = R' \underset{R_i}{\otimes} -$, and with canonical isomorphisms $\tau : \varphi_1\eta_1 \xrightarrow{\cong} \varphi_2\eta_2$.

Write $\underline{M}(R) = R_1\text{-mod} \underset{R'\text{-mod}}{\times} R_2\text{-mod}$, and let $U : R\text{-mod} \to \underline{M}(R)$ be the canonical functor $U(M) = (\eta_1 M, \tau_M, \eta_2 M)$. A result of [12] states that U has an adjoint functor S given by the following cartesian diagram :

$$\begin{array}{ccc} SM & \longrightarrow & M_2 \\ \downarrow & & \downarrow \\ M_1 & \to \varphi_1 M_1 \to & \varphi_2 M_2 \end{array}$$

Note that if M and N are modules over P" in the obvious sense, then $Q_\kappa(M \underset{P}{\times} N) = Q_\kappa(M) \underset{Q_\kappa(P)}{\times} Q_\kappa(N)$ so it follows that if $x = (M_1, \alpha, M_2)$ is in <u>Pic</u> then Sx is in (R,κ)-mod. More generally, let $\underline{M}(R,\kappa)$ be $(R_1,\kappa)\text{-mod} \underset{(R',\kappa)\text{-mod}}{\times} (R_2,\kappa)\text{-mod}$, with obvious structural morphisms, then we may extend the functor Q_κ to $\underline{M}(R)$ as follows : for $y = (M_1, \alpha, M_2)$, $Q_\kappa(y) = (Q_\kappa(M_1), \alpha_\kappa, Q_\kappa(M_2))$ where α_κ is the isomorphism given by :

$$\alpha_\kappa : Q_\kappa(Q_\kappa(M_1) \underset{R_1}{\otimes} R') = Q_\kappa(M_1 \underset{R_1}{\otimes} R') \xrightarrow{\cong} Q_\kappa(M_2 \underset{R_2}{\otimes} R') = Q_\kappa(Q_\kappa(M_2) \underset{R_2}{\otimes} R')$$

For $x \in \underline{M}(R,\kappa)$, $y \in \underline{M}(R)$ there is a functorial isomorphism :

$$\begin{array}{ccc} [Q_\kappa(y), x]_\kappa & \xrightarrow{\cong} & [y,x] \\ \| & & \| \\ \mathrm{Hom}_{\underline{M}(R,x)}(Q_\kappa(y), x) & & \mathrm{Hom}_{\underline{M}(R)}(y,x) \end{array}$$

Again writing T for the canonical functor as obtained before, we see that S is adjoint to T as follows : let $y \in (R,\kappa)$-mod, $x \in \underline{M}(R,\kappa)$ then we have $[Ty,x]_\kappa = [Q_\kappa(Uy),x]_\kappa = [Uy,x] = [y,Sx]$. Moreover, the canonical map $M \to SU(N)$ is an isomorphism for $N = R$, (see [12]) hence so is

$Q_\kappa(R) \to Q_\kappa(SU(R)) = SQ_\kappa(U(R)) = STQ_\kappa(R)$. Finally, as we have noted in the remarks preceding the theorem, T is cofinal. If we write $\perp$ for the product in both occurring categories then (by definition) we have for all M in Pic an N in Pic and a P in $\underline{\text{Pic}}(R,\kappa)$ such that $M \perp N \cong TP$. By assumption there exists a Q in $\underline{\text{Pic}}(R,\kappa)$ such that $P \perp Q \cong Q_\kappa(R)$, therefore :

$$M \perp (N \perp TQ) = TP \perp TQ \cong T(P \perp Q) \cong TQ_\kappa(R).$$

Furthermore : $SM \perp S(N \perp TQ) \cong STQ_\kappa(R) \cong Q_\kappa(R)$, and this implies that $SM \in \underline{\text{Pic}}(R,\kappa)$. The reader will easily verify that these functors define inverse equivalences. □

Given a diagram (*) and for some κ the deduced diagram (*'), we obtain a commutative exact diagram :

$$\begin{array}{ccccccccc} U(Q_\kappa(R)) & \to & U(Q_\kappa(R_1)) & \to & \text{Pic}(f_1,\kappa) & \to & \text{Pic}(R,\kappa) & \to & \text{Pic}(R_1,\kappa) \\ \downarrow & & \downarrow & & \downarrow\varphi & & \downarrow & & \\ U(Q_\kappa(R_2)) & \to & U(Q_\kappa(R')) & \to & \text{Pic}(f_2,\kappa) & \to & \text{Pic}(R_2,\kappa) & \to & \text{Pic}(R',\kappa) \end{array}$$

This allows to state :

II.20. COROLLARY. *Given a commutative diagram of commutative rings* (*), *then :*

1. *(excision). The canonical map* $\varphi : \text{Pic}(f_1,\kappa) \to \text{Pic}(f_1,\kappa)$ *is an isomorphism.*
2. *(Mayer-Vietoris). There is a long exact sequence :*

$$U(Q_\kappa(R)) \to U(Q_\kappa(R_1)) \oplus U(Q_\kappa(R_2)) \to U(Q_\kappa(R')) \to$$
$$\to \text{Pic}(R,\kappa) \to \text{Pic}(R_1,\kappa) \oplus \text{Pic}(R_2,\kappa) \to \text{Pic}(R',\kappa).$$

PROOF. Both results follow from the fact that (*) induces an E-surjective diagram (*') in which (F_1,F_2) is a cofinal pair of product preserving functors, by simple application of VII.6.1. and VII.4.3. in [12]. □

CHAPTER III
THE RELATIVE BRAUER GROUP

III.1. <u>κ-Progenerators and Morita Theory</u>.

Throughout this section R will be a commutative ring with unit and κ is an idempotent kernel functor on R-mod. An R-module E is <u>κ-quasiprojective</u> if the modules $Q_{R-P}(E)$ are projective in $Q_{R-P}(R)$-mod for all $P \in \underline{C}(\kappa)$. It is then clear that $Q_{R-P}(E)$ is $Q_{R-P}(R)$-projective even for $P \in K(\kappa)$, where $K(\kappa)$ is the complement of $L(\kappa)$ in Spec(R). If E' is a submodule of E such that E/E' is κ-torsion then E' is κ-quasiprojective if and only if E is κ-quasiprojective. In particular, if E is κ-torsion free then E is κ-quasiprojective if and only if $Q_\kappa(E)$ is.

III.1.1. <u>PROPOSITION</u>. *Any κ-quasiprojective* R-*module* E *is* κ-*flat.*

<u>PROOF</u>. We have to check that for any injective $i : M' \to M$ in R-mod the kernel of $E \otimes i : E \otimes M' \to E \otimes M$ is κ-torsion. Since $Q_{R-P}(E)$ is $Q_{R-P}(R)$-projective, hence certainly flat, it follows that $\mathrm{Ker}(Q_{R-P}(E) \otimes i) = 0$. Therefore $\mathrm{Ker}(E \otimes i)$ is κ_{R-P}-torsion for all $P \in C(\kappa)$, or $\mathrm{Ker}(E \otimes i)$ is κ-torsion since $\kappa = \inf_{P \in C(\kappa)} \{\kappa_{R-P}\}$.

III.1.2. <u>Note</u>. A κ-projective module is κ-quasiprojective.

An $E \in$ R-mod is said to be <u>κ-finitely generated</u> if there is a κ-open submodule E' of E such that E' is finitely generated (a κ-open submodule

X of Y will be one such that $\kappa(Y/X) = Y/X$). Obviously, the converse holds when R is Noetherian.

We say that E is <u>κ-finitely presented</u> if there is a finitely presented R-module E' and an R-linear $u : E' \to E$ such that Ker u and Coker u are κ-torsion. It is evident from the definition that u induces an isomorphism $Q_\kappa(E') \cong Q_\kappa(E)$. In case R is Noetherian E is κ-finitely presented if and only if it is κ-finitely generated. We say that E is <u>κ-faithful</u> if it is κ-closed and faithful as a $Q_\kappa(R)$-module. A κ-faithful $E \in R$-mod is a <u>κ-progenerator</u> if it is κ-quasiprojective and κ-finitely presented. Note that a κ-progenerator need not be a generator for R-mod nor for (R,κ)-mod !

III.1.3. <u>PROPOSITION</u>. *Let* $f : R \to S$ *be a morphism of commutative rings. If* E *is a* κ*-progenerator in* R*-mod then* $Q_\kappa(E \otimes_R S)$ *is a* κ*-progenerator in* S*-mod.*

<u>PROOF</u>. First we check that a κ-closed κ-finitely presented R-module E is a κ-progenerator if and only if each $Q_{R-P}(E)$ is a progenerator in $Q_{R-P}(R)$-mod for all $P \in C(\kappa)$. Clearly the κ-relative assumptions entail immediately that $Q_{R-P}(E)$ is finitely presented and projective in $Q_{R-P}(R)$-mod. If E' is a κ-open finitely generated submodule of E then $Q_{R-P}(E') = Q_{R-P}(E)$ for all $P \in C(\kappa)$. Consequently if $r \in Q_{R-P}(R)$ is such that $rQ_{R-P}(E) = 0$ then there is an $s \in R - P$ such that $sr \in j_{R-P}(R)$ and $sr\, j_{R-P}(E') = 0$. Hence $sr\,E' \subset \kappa_{R-P}(E')$. Now since E' is finitely generated in R-mod we may pick $t \in R-P$ such that $tsr\,E' = 0$. However, since E' is κ-open in E and E is κ-torsion free it follows that $tsr\,E = 0$ too; consequently $tsr = 0$ or $r \in \kappa_{R-P}(Q_{R-P}(R)) = 0$, proving that $Q_{R-P}(E)$ is also faithful. For the converse it suffices again to check that E is κ-faithful, which is easy enough. Having established our claim the proof is now straightforward. Indeed, $Q_\kappa(E \otimes_R S)$ is κ-closed. Moreover, since E is κ-finitely presented, we have an exact sequence :

$$0 \to \text{Ker } u \to E' \xrightarrow[u]{} E \to \text{Coker } u \to 0,$$

where both Ker u and Coker u are κ-torsion. Clearly Coker$(u \otimes S)$ is κ-torsion too.

Since

$$Q_\kappa(E' \otimes S) = Q_\kappa(Q_\kappa(E') \otimes S) \cong Q_\kappa(Q_\kappa(E) \otimes S) = Q_\kappa(E \otimes S)$$

we have that Ker$(u \otimes S)$ is κ-torsion. Finally, $E' \underset{R}{\otimes} S$ is finitely presented since E' is. It remains to check that $Q_{R-P}(Q_\kappa(E \underset{R}{\otimes} S))$ is a progenerator in $Q_{R-P}(S)$-mod for each P in $C_S(\kappa)$. Write $p = P \cap R$, $P \in C_S(\kappa)$, then certainly $p \in K(\kappa)$. We have :

$$Q_{S-P}(Q_\kappa(E \underset{S}{\otimes} S)) = Q_{S-P}(E \underset{R}{\otimes} S) = E \underset{R}{\otimes} S \underset{S}{\otimes} Q_{S-P}(S)$$

$$= E \underset{R}{\otimes} Q_{S-P}(S) = Q_{R-p}(E) \underset{Q_{R-p}(R)}{\otimes} Q_{S-P}(S)$$

(the latter because $\kappa_{R-p} \leqslant \kappa_{S-p}$ in S-mod !).

Since $Q_{R-p}(E)$ is a progenerator in $Q_{R-p}(R)$-mod the conclusion follows.

Let us extend notation of the foregoing chapter and write $M \perp_\kappa N$ or $M \perp N = Q_\kappa(M \underset{R}{\otimes} N)$, for each M,N in (R,κ)-mod.

III.1.4. <u>Remark</u>. Using the fact that $Q_\kappa(M \underset{R}{\otimes} N)$ equals $Q_\kappa(Q_\kappa(M) \underset{R}{\otimes} Q_\kappa(N))$ for each pair of R-modules M and N one easily verifies the following properties :

1. If M,N are κ-finitely generated (κ-finitely presented) then so is $M \perp N$.
2. $(M' \perp M) \perp N = M' \perp (M \perp N)$, for all M', M, N.
3. $Q_\kappa(R) \perp M = M$.
4. An R-bilinear map $M' \times M \to N$ where N is σ-closed factorizes uniquely through $M' \perp M$.

III.1.5. <u>LEMMA</u>. *If* $L \in R$-*mod is* κ-*flat and finitely generated then :*

$$\mathrm{Hom}_{Q_\kappa(R)}(Q_\kappa(L), Q_\kappa(L)) = Q_\kappa(\mathrm{Hom}_R(L,L)).$$

<u>PROOF</u>. Since L, hence $Q_\kappa(L)$, is κ-flat,

$$\mathrm{Hom}_{Q_\kappa(R)}(Q_\kappa(L),Q_\kappa(L)) = \mathrm{Hom}_R(Q_\kappa(L),Q_\kappa(L))$$

and the latter is a κ-closed R-module. Because $\mathrm{Hom}_R(Q_\kappa(L),Q_\kappa(L)) = \mathrm{Hom}_R(L,Q_\kappa(L))$ it will suffice to prove that Ker μ and Coker μ in the following exact sequence are κ-torsion,

$$0 \to \mathrm{Ker}\ \mu \to \mathrm{End}_R(L) \xrightarrow{\mu} \mathrm{Hom}_R(L,Q_\kappa(L)) \to \mathrm{Coker}\ \mu \to 0.$$

If $\varphi \in$ Ker μ and $j_\kappa : L \to Q_\kappa(L)$ is the localization morphism, then $j\varphi = 0$ means $\varphi(L) \subset \kappa(L)$. Since L is finitely generated there exists $I \in L(\kappa)$ such that $I\varphi(L) = 0$ i.e. $I\varphi = 0$ and thus $\varphi \in \kappa(\mathrm{End}_R(L))$. Consequently Ker μ is κ-torsion. Next, for $\varphi : L \to Q_\kappa(L)$ we have to check whether there exists a κ-open submodule L_1 in L together with a morphism $\varphi_1 : L_1 \to L$ making the following diagram commutative :

$$\begin{array}{ccc} L_1 & \longrightarrow & L \\ \downarrow{\scriptstyle \varphi_1} & & \downarrow{\scriptstyle \varphi} \\ L & \longrightarrow & Q_\kappa(L) \end{array}$$

However, if $L' = \varphi(L) \subset Q_\kappa(L)$ then there is an $I \in L(\kappa)$ such that $IL' \subset j_\kappa(L)$. Putting $L_1 = IL$ and $\varphi_1 = \varphi|L_1$ we obtain L_1 and φ_1 as desired. □

III.1.6. <u>COROLLARY</u>. *If* E *and* F *are κ-closed, κ-finitely presented κ-quasiprojective* R-*modules then :*

$$\mathrm{End}_R(E) \perp \mathrm{End}_R(F) = \mathrm{End}_R(E \perp F).$$

<u>PROOF</u>. By the lemma, it is clear that $\mathrm{End}_R(E)$ is κ-closed. If $u : E' \to E$ is R-linear with E' finitely presented and Ker u, Coker u are κ-torsion then $Q_\kappa(E') = Q_\kappa(E)$ and thus E' is κ-flat. Again from the lemma, $\mathrm{End}_R(E) = \mathrm{End}_R(Q_\kappa(E')) = Q_\kappa(\mathrm{End}_R(E'))$ follows. If F' is constructed in a similar way as E' then the canonical map $\mu : \mathrm{End}_R(F) \perp \mathrm{End}_R(F) \to \mathrm{End}_R(E \perp F)$ is given by $Q_\kappa(\mathrm{End}_R(E') \otimes_R \mathrm{End}_R(F')) \to Q_\kappa(\mathrm{End}_R(E' \otimes_R F'))$. For each $P \in C(\kappa)$ this induces an isomorphism :

$$\mathrm{End}_{Q_{R-P}(R)}(Q_{R-P}(E')) \otimes \mathrm{End}_{Q_{R-P}(R)}(Q_{R-P}(F')) \xrightarrow{\mu_P} \mathrm{End}_{Q_{R-P}(R)}(Q_{R-P}(E') \otimes Q_{R-P}(F'))$$

Indeed, E' is finitely presented, so $Q_{R-P}(\mathrm{End}_R(E')) = \mathrm{End}_{Q_{R-P}(R)}(Q_{R-P}(E'))$ and similarly for F' and $E' \otimes_R F'$. The fact that $Q_{R-P}(u)$ is an isomorphism for all $P \in C(\kappa)$ then implies that u is an isomorphism. □

III.1.7. Remark. Striving for mathematical economy in the foregoing one might think that κ-flatness for E and F would suffice to derive the result. However this niggardliness does not pay off because a κ-flat κ-finitely presented E is necessarily κ-quasiprojective.

PROOF. $Q_{R-P}(E)$ is finitely presented for each $P \in C(\kappa)$ and moreover $Q_{R-P}(E)$ is also flat in $Q_{R-P}(R)$-mod. A finitely presented flat module is projective. □

III.1.8. PROPOSITION. *If* E_1 *and* E_2 *are κ-progenerators then so is* $E_1 \perp E_2$.

PROOF. Easy. □

Let $E \in (R,\kappa)$-mod, then we write E^* for $\mathrm{Hom}_R(E,Q_\kappa(R))$ unless otherwise indicated.

III.1.9. PROPOSITION. *If* E *is a κ-progenerator then* $E^* \perp E$ *is isomorphic to* $\mathrm{End}_R(E)$.

PROOF. The canonical map $E^* \otimes_R E \to \mathrm{End}(E)_{Q_\kappa(R)}(E) = \mathrm{End}_R(E)$ factorizes through
$E^* \perp E \xrightarrow{\tau} \mathrm{End}_R(E)$ because $\mathrm{End}_R(E)$ is κ-closed. If $u : E' \to E$, with E' finitely presented, is an R-linear morphism with κ-torsion Ker and Coker then $E^* = \mathrm{Hom}_R(Q_\kappa(E'),Q_\kappa(R)) = \mathrm{Hom}_R(E',Q_\kappa(R))$. So for $P \in C(\kappa)$ we obtain :

$$Q_{R-P}(E^*) = Q_{R-P}(\mathrm{Hom}_R(E',Q_\kappa(R)) = \mathrm{Hom}_{Q_{R-P}(R)}(Q_{R-P}(E'),Q_{R-P}(R))$$

and the latter is the dual of $Q_{R-P}(E)$ in $Q_{R-P}(R)$-mod. Similarly, $Q_{R-P}(\mathrm{End}_R(E)) = \mathrm{End}_{Q_{R-P}(R)}(Q_{R-P}(E))$ for each $P \in C(\kappa)$. Now the fact that $Q_{R-P}(E)$ is a progenerator in $Q_{R-P}(R)$-mod entails that

$$\tau_P : Q_{R-P}(E^* \perp E) \to \mathrm{End}_{Q_{R-P}(R)}(Q_{R-P}(E)) = Q_{R-P}(\mathrm{End}_R(P))$$

is an isomorphism for each $P \in C(\kappa)$. Therefore τ is an isomorphism.

III.1.10. <u>Remark</u>. It is obvious that any κ-invertible R-module P is also κ-quasiprojective (because $Q_{R-Q}(P)$ is an invertible (hence free) $Q_{R-Q}(R)$-module for each $Q \in C(\kappa)$.) But on the other hand a κ-invertible R-module need not be a κ-progenerator because the constructions are not universal. In case R is Noetherian we do have a positive result :

III.1.11. <u>PROPOSITION</u>. *If* R *is Noetherian then the objects of* <u>Pic</u>(R,κ) *are κ-progenerators.*

<u>PROOF</u>. Consider P,Q in <u>Pic</u>(R,κ) such that $Q_\kappa(P \otimes Q) \cong Q_\kappa(R)$. For all $p \in C(\kappa)$ we obtain that $Q_{R-p}(P) \otimes Q_{R-p}(Q) \cong Q_{R-p}(R)$ hence $Q_{R-p}(P)$ is an invertible $Q_{R-p}(R)$-module i.e. also free ! That P is κ-quasiprojective is evident. Since R is Noetherian we only have to check whether P is κ-finitely generated. From $Q_\kappa(P \otimes Q) \cong Q_\kappa(R)$ it is obvious that $P \otimes Q$ is κ-finitely generated, let T be finitely generated and κ-open in $P \otimes Q$. Actually we may assume that T is generated by some collection

$\{p_i \otimes q_i;\ i = 1,\dots,m,\ p_i \in P,\ q_i \in Q\}$. If we write
$P_1 = Rp_1 + \dots + Rp_m \subset P$ then we get :

$$0 \to K \to P_1 \otimes Q \xrightarrow{\phi} P \otimes Q,$$

where K is κ-torsion. Clearly $T \subset \mathrm{Im}\ \phi$ and so we obtain inclusions $Q_\kappa(T) \hookrightarrow Q_\kappa(\mathrm{Im}\ \phi) \hookrightarrow Q_\kappa(P \otimes Q)$ where the composition is an isomorphism by the choice of T. It is clear that $Q_\kappa(\mathrm{Im}\ \phi) \cong Q_\kappa(P \otimes Q)$ and $Q_\kappa(\mathrm{Im}\ \phi) \cong Q_\kappa(P_1 \otimes Q)$, therefore we may identify $Q_\kappa(P \otimes Q)$ and $Q_\kappa(P_1 \otimes Q)$.

Now,

$$Q_\kappa(P_1) \cong Q_\kappa(P_1) \perp Q_\kappa(R) = Q_\kappa(P_1) \perp (Q \perp P)$$
$$= (Q_\kappa(P_1) \perp Q) \perp P = Q_\kappa(P_1 \perp Q) \perp P = Q_\kappa(R) \perp P = P.$$

To check that P is κ-faithful apply III.1.4. □

III.1.12. <u>COROLLARY</u>. *If* $P \in R$-*mod be* κ-*invertible, then* $Q_\kappa(P \otimes [P, Q_\kappa(R)]) = Q_\kappa(R)$. □

We conclude this section with some remarks concerning a relativation of the concept of Morita equivalence and related topics.

Is S is a commutative R-algebra and $M,N \in S$-mod then we write $M \underset{S}{\perp} N = Q_\kappa(M \underset{S}{\otimes} N)$ and this is obviously an S-module. If E is a κ-progenerator then we have seen that $E^* \perp E = \mathrm{End}_R(E)$. Moreover it is clear that the canonical map $E^* \underset{B}{\otimes} E \to Q_\kappa(R)$ localizes at each $P \in C(\kappa)$ to an isomorphism and therefore : $E^* \underset{B}{\perp} E \cong Q_\kappa(R)$ (Behold : $Q_{R-P}(E^*) = (Q_{R-P}(E))^*$ for all $P \in C(\kappa)$!)

Write (B,R,κ)-mod for the full subcategory of B-mod consisting of left B-modules which are κ-closed as an R-module. Let us denote the kernel functor induced by κ in B-mod by κ again. Obviously (B,R,κ)-mod is a full subcategory of (B,κ)-mod.

III.1.13. <u>PROPOSITION</u>. *Let* E *be a* κ-*progenerator for* R-*mod and write* $B = \mathrm{End}_R(E)$ *then the functor* $- \perp E : (R,\kappa)\text{-mod} \to (B,R,\kappa)\text{-mod}$ *is an equivalence of categories with inverse equivalence given by* $E^* \underset{B}{\perp} -$.

<u>PROOF</u>. If $L \in (R,\kappa)$-mod, then we have :

$$E^* \underset{B}{\perp} (L \perp E) = E^* \underset{B}{\perp} (E \perp L) = (E^* \underset{B}{\perp} E) \perp L = R \perp L = L.$$

(Since R is central in B the notation $\underset{B}{\perp}$ makes sense here).
On the other hand for any B-module M in (B,R,κ)-mod we obtain :

$(E^* \perp M) \perp E = (M \underset{B^{opp}}{\perp} E^*) \perp E = M \underset{B^{opp}}{\perp} (E^* \perp E) = M \underset{B^{opp}}{\perp} B = B \underset{B}{\perp} M = M.$

The proposition is therefore easily verified. □

III.1.14. LEMMA. *Let* R *be a Noetherian ring and let* $M \in$ R-mod *be* κ*-finitely generated, then* M^* *is* κ*-finitely generated too. If* R *is* κ*-closed then* M^* *is finitely generated.*

PROOF. By definition there is a κ-open submodule M' in M, which is finitely generated. Let, for some n, $\pi : R^n \to M$ be a surjective morphism. The map π induces an injection $\pi^* : M^* = (M')^* \hookrightarrow \mathrm{Hom}_R(R^n, Q_\kappa(R)) = Q_\kappa(R)^n$. Put $N = (\pi^*)^{-1}(j_\kappa(R)^n)$. Since $j_\kappa(R)^n$ is κ-open in $Q_\kappa(R)^n$ it follows that N is κ-open in M^*, but as N is evidently finitely generated this proves that M^* is κ-finitely generated. If $R = Q_\kappa(R)$ holds then $\pi^* : M^* \subset R^n$ yields that M^* is finitely generated. □

III.1.15. LEMMA. *Let* $M \in (R,\kappa)$*-mod be* κ*-finitely generated then* M *is also* κ*-finitely generated as a* $Q_\kappa(R)$*-module.*

PROOF. If $N \subset M$ is finitely generated and κ-open in M then let $\pi : R^n \to N$ be a surjective morphism for some $n \in \mathbb{N}$. Localization at κ yields a $Q_\kappa(R)$-linear map $Q_\kappa(\pi) : Q_\kappa(R)^n \to Q_\kappa(N) = Q_\kappa(M) = M$ which need not be surjective. However, Im $Q_\kappa(\pi)$ is finitely generated, and κ-open in M because it contains N. □

III.1.16. LEMMA. *Assume* $Q_\kappa(R)$ *is Noetherian. If* M *is* κ*-finitely generated,* κ*-closed and* κ*-quasiprojective in* R-mod *then* M *is a reflexive* $Q_\kappa(R)$*-module.*

PROOF. Since $Q_\kappa(R)$ is Noetherian it follows from foregoing lemmas that $M^* = \mathrm{Hom}_{Q_\kappa(R)}(M, Q_\kappa(R))$ is finitely generated and thus also finitely presented as a $Q_\kappa(R)$-module. So for all $P \in C(\kappa)$ we have that :

$$Q_{R-P}(M^{**}) = Q_{R-P}(\mathrm{Hom}_{Q_\kappa(R)}(M^*, Q_\kappa(R))) = \mathrm{Hom}_{Q_{R-P}(R)}(Q_{R-P}(M^*), Q_{R-P}(R)).$$

But M is κ-finitely generated, hence κ-finitely generated as a $Q_\kappa(R)$-module, i.e. there is a finitely presented submodule N in M which is κ-open in M. Therefore :

$$Q_{R-P}(M^*) = Q_{R-P}(\mathrm{Hom}_{Q_\kappa(R)}(M, Q_\kappa(R))) = Q_{R-P}(\mathrm{Hom}_{Q_\kappa(R)}(M_1, Q_\kappa(R)))$$

$$= \mathrm{Hom}_{Q_{R-P}(R)}(Q_{R-P}(M_1), Q_{R-P}(R)) = (Q_{R-P}(M))^*.$$

From our assumptions it then follows that $Q_{R-P}(M^{**})$ equals $(Q_{R-P}(M))^{**} = Q_{R-P}(M)$. This holds for all $P \in C(\kappa)$ and thus it follows that $Q_\kappa(M^{**}) = Q_\kappa(M) = M$. Any κ-quasiprojective R-module is necessarily κ-flat and so it follows from foregoing results that $M^{**} = Q_\sigma(M^{**})$, and combining this with the above we find $M = M^{**}$ or M is reflexive in $Q_\kappa(R)$-mod. □

III.1.17. COROLLARY. *Consider a Noetherian ring and κ-progenerators* E *and* F *for* R-*mod. Write* $B = \mathrm{End}_R(E)$ *and assume that* F *is a* B-*module on the left. Then* $S = \mathrm{Hom}_B(E,F)$ *is a κ-progenerator for* R-*mod.*

PROOF. From Proposition III.1.13. it follows that

$$S = \mathrm{Hom}_R(E^* \underset{B}{\perp} E, E^* \underset{B}{\perp} F) = \mathrm{Hom}_R(R, E^* \underset{B}{\perp} F) = E^* \underset{B}{\perp} F.$$

In particular, S is κ-closed, so we only have to establish that $E^* \underset{B}{\perp} F$ is κ-finitely generated. Since R is Noetherian it will suffice to verify that $E^* \underset{B}{\otimes} F$ is κ-finitely generated. The latter is an epimorphic image of $E^* \otimes F$ hence we have to check whether $E^* \underset{R}{\otimes} F$ or $E^* \perp F$ is κ-finitely generated, but this is easily seen from Remark III.1.4. (1) and the foregoing lemma. □

III.2. Separability.

Throughout this section R is a commutative ring and A is an R-algebra. Notations and conventions are as in Section I.3. The multiplication map $m_A : A^e \to A$ is left A^e-linear and if A is commutative then m_A is a ring homomorphism. Write $J = \mathrm{Ker}\, m_A$, so we obtain the following exact sequence in A^e-mod :

$$(*) \qquad 0 \to J \to A^e \xrightarrow{m_A} A \to 0.$$

It is clear that J is exactly the left ideal of A^e generated by the set $\{1 \otimes a - a \otimes 1;\ a \in A\}$.

The algebra A is a separable R-algebra if (*) is a split exact sequence, see I.3. A left and right A-module M centralizing the action of R will be called an A-R-module. The category of A-R-modules is equivalent to A^e-mod, so usually we shall talk about A^e-modules. If $M \in A^e$-mod, put $M^{(A)} = \{m \in M;\ am = ma \text{ for all } a \in A\}$. The operator $(-)^{(A)}$ defines a covariant functor from A^e-mod → R-mod. Let us recall some basic properties in the following :

III.2.1. <u>LEMMA</u>. *Let* A *be an* R-*algebra.*

1. *If* $M \in A^e$-mod, *then* $M^{(A)} \cong \mathrm{Hom}_{A^e}(A,M)$ *in* R-mod.
2. $\mathrm{Hom}_{A^e}(A,A) \cong Z(A)$, *the center of* A.
3. *The right annihilator of* J *in* A^e *is just* $\mathrm{Hom}_{A^e}(A,A^e)$ *and if* A *is a separable* R-*algebra then we have that* $m_A(\mathrm{Hom}_{A^e}(A,A^e)) \cong Z(A)$.
4. A *is separable over* R *if and only if the functor* $(-)^{(A)}$ *is right exact.*
5. *Consider commutative* R-*algebras* C_1, C_2 *and let* A_1 *be a separable* C_1-*algebra,* A_2 *a separable* C_2-*algebra. If* $A_1 \otimes_R A_2 \neq 0$, $C_1 \otimes_R C_2 \neq 0$ *then* $A_1 \otimes_R A_2$ *is a separable* $C_1 \otimes_R C_2$-*algebra.*
6. *As in* 5. *but with* A_2 *being an arbitrary* R-*algebra. If* A_2 *is a faithful* R-*module and* A_2 *contains* R *as a direct summand as an* R-*module, then* $C_1 \otimes_R C_2$-*separability of* $A_1 \otimes_R A_2$ *entails* C_1-*separability of* A_1.
7. *Let* C_1 *be a commutative* R-*algebra containing* R *as a direct summand as an* R-*module. Then* A *is a separable* R-*algebra if and only if* $A_1 \otimes_R C_1$ *is a separable* C_1-*algebra. If the image of* $1 \otimes C_1$ *is the center of* $A \otimes C_1$ *then* R *is the center of* A.
8. *Let* C_1 *be a commutative separable* R-*algebra and let* A *be a separable* C_1-*algebra, then* A *is a separable* R-*algebra. Furthermore if* A *is* R-*separable while* C_1 *is a separable* R-*algebra contained in* Z(A) *then* A *is a separable* C_1-*algebra.*

An <u>extension of R</u> will be a commutative faithful R-algebra. We always assume that R is a subring of the extensions considered.
If C is an extension of R and if G is a group of R-automorphisms of C then C is said to be a <u>normal extension of R, if</u> $R = C^G = \{x \in C;\ \sigma(x) = x,$ for all $\sigma \in G\}$.

It is possible to state an equivalent for the fundamental theorem of Galois theory for fields, but we refer to [20] for this and related results.

III.2.2. <u>DEFINITION</u>. Let C be an extension of R and let G be a finite group of R-automorphisms of C. We say that C is a <u>Galois extension of R</u> with <u>Galois group G</u> if one of the following equivalent conditions holds :

G1. a. $C^G = R$.

b. For any idempotent $e \neq 0$ in C and each pair $\sigma,\tau \in G$ with $\sigma \neq \tau$, there is an $x \in C$ such that $\sigma(x)e \neq \tau(x)e$.

c. C is a separable R-algebra.

G2. a. $C^G = R$.

b. There exist $x_1,\ldots,x_n$, $y_1,\ldots,y_n$ in C such that $\sum_{j=1}^{n} x_j\sigma(y_j) = \delta(\sigma,1_G)$, for each $\sigma \in G$.

G3. a. C is a finitely generated projective R-module.

b. $\mathrm{End}_R(C)$ is isomorphic as an R-algebra to the crossed product algebra $A(C,G,1)$ obtained by defining multiplication on the free C-module $C\{u_\sigma;\sigma \in G\}$ by putting $(C_1u_\sigma)(C_2u_\tau) = C_1\sigma(C_2)u_{\sigma\tau}$ (extended linearly).

G4. a. $C^G = R$.

b. $C \otimes_R C$ is isomorphic as an R-algebra to the direct sum of $|G|$ copies of C.

G5. a. $C^G = R$.

b. For each maximal ideal P of C and for each $\sigma \neq 1$ in G there is an $x \in C$ such that $\sigma(x) - x \notin P$.

III.2.3. Remark. The isomorphism mentioned in G.3. b., say $j : A(C,G,1) \to End_R(C)$, may be defined by $(j(du_\sigma))(x) = d\sigma(x)$, for any $d,x \in C$, $\sigma \in G$. We define $tr \in Hom_R(C,R) = C^*$ by $tr = j(\sum_{\sigma\in G} u_\sigma)$. If C is a Galois extension of R with Galois group G then tr is a free generator for $Hom_R(C,R)$ viewed as a C-module.

Finally we list some properties of Galois extensions, again referring to [20], [33] or [68] for full detail.

II.2.4. PROPOSITION. *Let* C *be a Galois extension of* R *with Galois group* G.

1. R *is a direct summand of* C *in* R-*mod.*
2. C *is constant rank in* R-*mod equal to* $|G|$.
3. *For any commutative* R-*algebra* D, $C \otimes_R D$ *is a Galois extension of* D *with Galois group* G.
4. *If* C *does not contain nontrivial idempotents then any* R-*algebra morphism* $\gamma : C \to C$ *is in* G.
5. *There exists a* $c \in C$ *such that* $tr(c) = 1$. □

Now we return to the situation of the exact sequence (*) and we assume moreover that both R and A are κ-closed with respect to some kernel functor κ on R-mod. Localizing (*) at κ yields :

$$0 \to Q_\kappa(J) \to Q_\kappa(A^e) \xrightarrow{m_A} A \to 0$$

where $m_A = Q_\kappa(m_A)$. We say that A is a κ-separable R-algebra if A is a κ-quasiprojective $Q_\kappa(A^e)$-algebra. It is clear how the local equivalents (at each $P \in C(\kappa)$) of statements in III.2.1. may be phrased. In particular the functor $(-)^A$ restricts to a functor from κ-closed left A^e-modules to (A^A,κ)-mod which need not be exact but which is exact at each $P \in C(\kappa)$ in the obvious sense.

III.2.5. DEFINITION. A commutative R-algebra S is said to be a κ-Galois extension of R if :

κG.1. S is a κ-separable extension of R and a κ-progenerator.

κG.2. There is a finite group of R-automorphisms of S, G say, such that : $R = S^G$.

κG.3. Let $\bar{\kappa}$ denote the kernel functor on S-mod induced by κ, then for each $P \in C(\bar{\kappa})$ and each $\sigma \neq 1$ in G there exists an $s \in S$ such that $\sigma(s) - s \notin P$.

III.2.6. PROPOSITION. *Let* S *be a commutative* R*-algebra satisfying conditions* κG.1, κG.2 *above, then the following statements are equivalent :*

1. S *is a* κ*-Galois extension of* R.
2. *For each* $P \in C(\kappa)$, $Q_{R-P}(S)$ *is a Galois extension of* $Q_{R-P}(R)$ *with Galois group* G.

PROOF. From κG.1 and κG.2 it follows that $Q_{R-P}(S)$ is a separable $Q_{R-P}(R)$-algebra and finitely generated (free) as a $Q_{R-P}(R)$-module too. The maximal ideals of $Q_{R-P}(S)$ correspond bijectively to the $P \in C(\bar{\kappa})$. The implication 1. ⇒ 2. follows directly from G.5 in Definition III.2.2. applied to $Q_{R-P}(S)$. The converse implication also follows from G.5 because if $\bar{x} \in Q_{R-P}(S)$ is such that $\sigma(\bar{x}) - \bar{x} \notin Q_{R-P}(P)$ for $P \in C(\bar{\kappa})$ then there is a $c \in R - P$ such that $c\bar{x} \in j_{R-P}(S)$ and $\sigma(c\bar{x}) - c\bar{x} = c(\sigma(\bar{x}) - \bar{x}) \notin j_{R-P}(P)$. If $y \in S$ represents cx then $\sigma(y) - y \notin P$. □

III.2.6. COROLLARY. *Condition* 2 *above is equivalent to* 2' *where* $C(\kappa)$ *is replaced by* $K(\kappa)$. □

III.3. The Relative Brauer Group.

Let R be a commutative ring and A an R-algebra.

III.3.1. LEMMA.

1. *For each kernel functor* κ *on* R*-mod,* $Q_\kappa(A)$ *has an essentially unique* R*-algebra structure extending the structure of* A.
2. *If* $\{\kappa_i; i \in I\}$ *is a family of idempotent kernel functors on* R*-mod and*

$\kappa = \inf\{\kappa_i; i \in I\}$ *then* $Q_\kappa(A)$ *is a central* $Q_\kappa(R)$*-algebra if* $Q_{\kappa_i}(A)$ *is a central* $Q_{\kappa_i}(R)$*-algebra for each* $i \in I$.

PROOF. 1. Is well-known and in fact we have used it implicitly throughout foregoing sections.

2. From $Z(A) \subset A$, $Q_\kappa(Z(A)) \subset Q_\kappa(A)$ and $Q_\kappa(Z(A)) \subset Z(Q_\kappa(A))$ follows. Indeed, let us write Z for $Z(A)$, then we have a commutative diagram :

$$\begin{array}{ccccccccc} 0 & \to & \kappa(Z) & \to & Z & \to & \overline{Z} & \to & 0 \\ & & \downarrow & & \downarrow & & \downarrow & & \\ 0 & \to & \kappa(A) & \to & A & \to & \overline{A} & \to & 0 \end{array}$$

and vertical arrows represent monomorphisms.
Clearly $\overline{Z} \subset Z(\overline{A})$. If $x \in Q_\kappa(A)$ and $z \in Q_\kappa(Z)$ then for some $I,J \in L(\kappa)$, $Ix \subset \overline{A}$, $Jz \in \overline{Z}$. Therefore : $IJ(xz-zx) = 0$ i.e. $xz - zx \in \kappa(Q_\kappa(A)) = 0$ and thus $Q_\kappa(Z) \subset Z(Q_\kappa(A))$ follows. Consider the following diagram of ring morphisms :

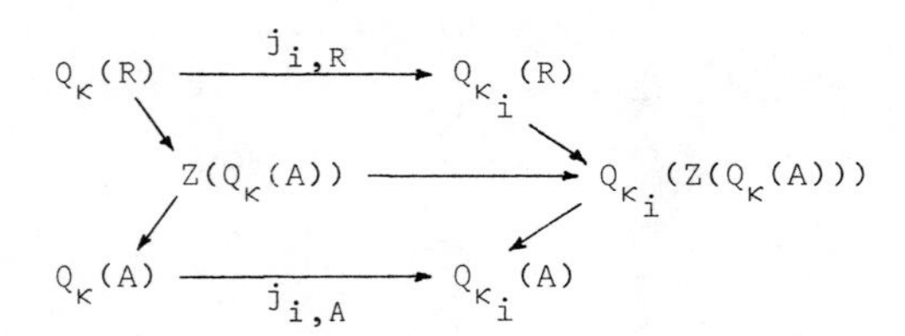

Evidently, $Q_\kappa(R) \subset Q_\kappa(Z(A)) \subset Z(Q_\kappa(A))$ and $j_{i,A}$ maps $Z(Q_\kappa(A))$ into $Z(Q_{\kappa_i}(A)) = Q_{\kappa_i}(R)$. On the other hand we have :

$$Q_{\kappa_i}(Z(Q_\kappa(A))) \subset Z(Q_{\kappa_i}(Q_\kappa(A))) = Z(Q_{\kappa_i}(A)) = Q_{\kappa_i}(R).$$

Hence, for each $i \in I$ we obtain :

$$Q_{\kappa_i}(Z(Q_\kappa(A))) = Q_{\kappa_i}(R) = Q_{\kappa_i}(Q_\kappa(R)).$$

It follows that $Z(Q_\kappa(A))/Q_\kappa(R)$ is κ_i-torsion for all $i \in I$, hence it is also a κ-torsion R-module. But then

$$Z(Q_\kappa(A)) \subset Q_\kappa(Z(Q_\kappa(A))) = Q_\kappa(R),$$

i.e. $Z(Q_\kappa(A)) = Q_\kappa(R)$. Note that the first inclusion holds because $Z(Q_\kappa(A))$ is κ-torsion free. □

In the following we use some results about P.I.-algebras for which we refer to [7], [71], ...

III.3.2. THEOREM. *If* A *is a* κ*-torsion free* R*-algebra, for some idempotent kernel functor* $\kappa = \inf\{\kappa_{R-P}; P \in C(\kappa)\}$ *on* R*-mod, and if* A *is* κ*-finitely generated as an* R*-module then* A *is a P.I. algebra.*

PROOF. For each $P \in C(\kappa)$ we have that $Q_{R-P}(A)$ is a finitely generated $Q_{R-P}(R)$-module, consequently $Q_{R-P}(A)$ is a PI-algebra and actually it satisfies the standard identity S_{1+n}, cf. [71], where n is the number of generators for some finitely generated R-module which is κ-open in A. Now we have :

$$0 = \kappa(A) = \bigcap_{P \in C(\kappa)} \kappa_{R-P}(A)$$

and therefore there is a monomorphic embedding

$$0 \to A \to \prod_{P \in C(\kappa)} A/\kappa_{R-P}(A) = B.$$

The ring $A/\kappa_{R-P}(A)$ satisfy the identity S_{1+n}, hence B and a fortiori A satisfies S_{1+n} too. Then A is a P.I.-algebra. □

III.3.3. COROLLARY. *If* A *is a* κ*-closed central* R*-algebra which is* κ*-finitely generated as an* R*-module and if we assume that* R *is a Noetherian ring then* A *is a finitely generated* R*-module.*

PROOF. By the theorem A is a PI-algebra with Noetherian center. A result of G. Cauchon, yields that A is a finitely generated R-module. □

III.3.4. COROLLARY. *If* A *is finitely generated as an algebra over some Noetherian subring of* R *and if* A *is* κ*-closed* R*-central while* R *is a do-*

main such that its integral closure is a Krull domain, then A *is* κ*-finitely generated if and only if it is finitely generated.*

PROOF. By the theorem A is a PI-algebra and it is affine because of the hypothesis. The statement is then a direct consequence of a result of A. Braun. □

III.3.5. COROLLARY. *A* κ*-separable* κ*-finitely generated* R*-central algebra over a Noetherian ring* R *is a finite* R*-module and also a Noetherian ring.*

III.3.6. COROLLARY. *Let* R *be a Noetherian ring and let* A *be a* κ*-torsion free central* R*-algebra which is* κ*-finitely generated as an* R*-module then for any kernel functor* $\sigma \geq \kappa$, $Q_\sigma(A)$ *is a central* $Q_\sigma(R)$*-algebra.*

PROOF. That $Q_\sigma(R)$ is in the center of $Q_\sigma(A)$ is obvious. On the other hand if $z \in Z(Q_\sigma(A))$ then for some $I \in L(\sigma)$, $Iz \subset A/\sigma(A)$, thus $Iz \subset Z(A/\sigma(A))$. For each $\overline{x} \in Z(A/\sigma(A))$, $x \in A$, it follows that $xy - yx \in \sigma(A)$ for all $y \in A$. By the theorem A is a finitely generated R-module hence $\sigma(A)$ is a finitely generated R-module. Therefore $J\sigma(A) = 0$ for some $J \in L(\sigma)$. It follows that $Jx \subset Z(A) = R$ i.e. $J\overline{x} \in R/\sigma(R)$ i.e. $\overline{x} \in Q_\sigma(R)$. Since $Iz \subset Z(A/\sigma(A))$ it follows that $JIZ \subset R/\sigma(R)$ and thus from $JI \in L(\sigma)$, $Z \in Q_\sigma(R)$ follows. □

III.3.7. DEFINITION. An R-algebra A is a κ-Azumaya algebra if it is a κ-progenerator such that the canonical map $A^e \to \mathrm{End}_R(A)$ induces an isomorphism $Q_\kappa(A^e) \cong \mathrm{End}_R(A)$.

III.3.8. PROPOSITION. *Let* A *be a* κ*-closed* R*-algebra which is* κ*-finitely presented as an* R*-module then the following statements are equivalent :*

1. A *is a* κ*-Azumaya algebra*
2. $Q_{R-P}(A)$ *is an Azumaya algebra over* $Q_{R-P}(R)$ *for all* $P \in C(\kappa)$.

PROOF. 1. ⇒ 2. Immediate from $Q_{R-P}(A^e) = (Q_{R-P}(A))^e$ and

$Q_{R-P}(\mathrm{End}_R(A)) = \mathrm{End}_{Q_{R-P}(R)}(Q_{R-P}(A))$, combined with Theorem I.3.4.

2. ⇒ 1. A κ-closed κ-finitely presented R-module E is a κ-progenerator if and only if each $Q_{R-P}(E)$ is a progenerator in $Q_{R-P}(R)$-mod for all $P \in C(\kappa)$. This has been proved in the proof of Proposition III.1.3. Since both $Q_\kappa(A^e)$ and $\mathrm{End}_R(A)$ are κ-closed (see Proposition II.3.) the isomorphism $Q_\kappa(A^e) \cong \mathrm{End}_R(A)$ derives directly from the given local isomorphisms. □

III.3.9. Remark. If A is a κ-Azumaya R-algebra then $Z(A) = Q_\kappa(R)$ follows from III.3.8. and Lemma III.3.1.

III.3.10. PROPOSITION. *If* E *is a* κ-*progenerator then* $\mathrm{End}_R(E)$ *is a* κ-*Azumaya algebra.*

PROOF. If E is a κ-progenerator then $Q_{R-P}(E)$ is a progenerator in $Q_{R-P}(R)$-mod for all $P \in C(\kappa)$ and $\mathrm{End}_{Q_{R-P}(R)}(Q_{R-P}(E))$ is an Azumaya algebra. But $Q_{R-P}(\mathrm{End}_R(E)) = Q_{R-P}(\mathrm{End}_R(Q_\kappa(E')))$ for some finitely presented R-module E', κ-open in E. Then also :

$$Q_{R-P}(\mathrm{End}_R(E)) = Q_{R-P}(Q_\kappa(\mathrm{End}_R(E'))) = \mathrm{End}_{Q_{R-P}(R)}(Q_{R-P}(E'))$$

$$= \mathrm{End}_{Q_{R-P}(R)}(Q_{R-P}(E)).$$

Applying Proposition III.3.8. finishes the proof. □

We introduce the notation $A \perp B = Q_\kappa(A \underset{R}{\otimes} B)$ for R-algebras A and B; if we want to specify κ or R we may somtimes use $\underset{\kappa}{\perp}$, $\underset{R}{\perp}$.

III.3.11. PROPOSITION. *If* A_1 *and* A_2 *are* κ-*Azumaya algebras over* R *then so is* $A_1 \perp A_2$.

PROOF. Straightforward (using Proposition III.3.8.). □

Let A_1, A_2 be κ-Azumaya algebras over R. We say that A_1 is <u>κ-similar</u> or <u>κ-equivalent</u> to A_2 if there exist κ-progenerators E_1, E_2 such that $A_1 \perp \mathrm{End}_R(E_1) \cong A_2 \perp \mathrm{End}_R(E_2)$. It is easily checked that κ-similarity is an equivalence relation. The κ-similarity class of a κ-Azumaya algebra A over R will be denoted by [A] and we write $\mathrm{Br}(R,\kappa)$ for the set of κ-similarity classes of κ-Azumaya algebras over R. We can introduce a product operation in $\mathrm{Br}(R,\kappa)$ by putting $[A_1].[A_2] = [Q_\kappa(A_1 \otimes_R A_2)]$. The unit element for this multiplication is given by $[\mathrm{End}_R(E)]$ for some κ-progenerator E.

From Remark III.1.4., Corollary III.1.6. and Proposition III.1.8. one easily deduces that $\mathrm{Br}(R,\kappa)$ is an abelian group, and in particular we have $[A]^{-1} = [A^0]$. The group $\mathrm{Br}(R,\kappa)$ is said to be the <u>relative Brauer group with respect to κ</u>, or shortly, the <u>κ-Brauer group</u>.

III.3.12. <u>PROPOSITION</u>. *Let* R *be a Noetherian ring and let* A_1, A_2 *be* κ-*Azumaya algebras over* R. *The following properties are equivalent :*

1. A_1 *is* κ-*similar to* A_2
2. *There is a* κ-*progenerator* E *such that*

$$Q_\kappa(A_1 \otimes_R A_2^0) \cong A_1 \perp A_2^0 \cong \mathrm{End}_R(E).$$

<u>PROOF</u>. Evidently we only have to establish the following. A class [A] in $\mathrm{Br}(R,\kappa)$ is the unit class if and only if we may find a κ-progenerator E and an isomorphism $A \cong \mathrm{End}_R(E)$. By definition there exist κ-progenerators E and F such that $A \perp \mathrm{End}_R(E) \cong \mathrm{End}_R(F)$. Put $B = \mathrm{End}_R(E)$. The canonical mapping

$$A \otimes_R B \to Q_\kappa(A \otimes_R B) = A \perp B \cong \mathrm{End}_R(F)$$

endows F with a B-module structure, on the left say. Put $S = \mathrm{Hom}_B(E,F)$. Corollary III.1.17. entails that S is a κ-progenerator and $S = E^* \perp_B F$. Moreover, $S \perp E \cong F$ is obvious. From Proposition III.1.13. it follows that $\mathrm{End}_R(S) = \mathrm{End}_B(S \perp E) = \mathrm{End}_B(F)$. Verification of the isomorphism $\mathrm{End}_B(F) \cong A$ may now be carried out along the lines of the similar result for Azumaya algebras, e.g. as in [33]; we leave these details to the reader. □

A calculation of the relative Brauer group is sometimes rather easy, as the following proposition amply displays.

III.3.13. PROPOSITION. *Let* R *be a Noetherian domain which is* κ*-closed for some* $\kappa = \inf\{\kappa_{R-P}; P \in C(\kappa)\}$. *Assume that* $Q_{R-P}(R)$ *is a discrete valuation ring for all* $P \in C(\kappa)$, *then* $Br(R,\kappa) = \bigcap_{P\in C(\kappa)} Br(Q_{R-P}(R))$.

PROOF. The field of fractions of R will be denoted by K. As a first step we prove that the canonical maps $A \to A \otimes_R K$ define an injective group homomorphism $Br(R,\kappa) \to Br(K)$. Indeed, if A is a κ-Azumaya algebra over R then A is a prime PI-ring by Theorem III.3.2. and therefore $\Sigma = A \otimes_R K$ is a central simple algebra over K. If Σ represents the unit of Br(K) then $\Sigma = End_K(V)$ for some finite dimensional K-vectorspace V. Let $\{v_1,\ldots,v_n\}$ be a K-basis for V, write $F = Rv_1 \oplus\ldots\oplus Rv_n$ and $E_1 = AF$. From Corollary III.3.5. it follows that A is a finitely generated R-module, hence so is E_1. If $E = Q_\kappa(E_1^{**})$ then there are canonical inclusions $A \subset End_R(E_1) \subset End_R(E_1^{**}) \subset End_R(E)$. In order to check that E is a κ-progenerator it will suffice to check that E is κ-quasiprojective. For each $P \in C(\kappa)$, $Q_{R-P}(E) = Q_{R-P}(E_1^{**}) = Q_{R-P}(E)^{**}$ and therefore $Q_{R-P}(E)$ is a finitely generated reflexive $Q_{R-P}(R)$-module, hence projective (hence free) because $Q_{R-P}(R)$ is a discrete valuation ring, cf. [36]. Put $B = End_R(E)$, then B is a κ-Azumaya algebra by the foregoing. So we obtain Azumaya algebras $Q_{R-P}(A)$ and $Q_{R-P}(B)$ over $Q_{R-P}(R)$, hence since $B \subset End_K V = \Sigma$, it follows that $Q_{R-P}(A) = Q_{R-P}(B)$ because both are maximal orders of Σ over $Q_{R-P}(R)$.
Consequently, $A = B = End_R(E)$ or [A] is trivial in $Br(R,\kappa)$. So far we have established that $Br(R,\kappa) \subset \bigcap_{P\in C(\kappa)} Br(Q_{R-P}(R)) \subset Br(K)$, the second inclusion deriving from the fact that $Br(Q_{R-P}(R)) \subset Br(K)$. For the converse inclusion $G = \bigcap_{P\in C(\kappa)} Br(Q_{R-P}(R)) \subset Br(R,\kappa)$, we may use the theory of maximal orders of Σ over integrally closed Noetherian domains. Indeed, R is integrally closed since $R = \bigcap_{P\in C(\kappa)} Q_{R-P}(R)$!
A class in G may be represented by a skewfield Δ over K with the proper-

ty that for all $P \in C(\kappa)$ there is an Azumaya algebra $A(P)$ over $Q_{R-P}(R)$ such that $A(P) \underset{Q_{R-P}(R)}{\otimes} K$ is similar to Δ.
Consider any maximal order A of Δ over R. Then $Q_{R-P}(A)$ is a maximal order in Δ over $Q_{R-P}(R)$, and as $Q_{R-P}(R)$-orders are conjugated it follows that $Q_{R-P}(A) \cong A(P)$. Therefore, for all $P \in C(\kappa)$, A localizes to an Azumaya algebra $Q_{R-P}(A)$, i.e. A is a κ-Azumaya algebra over R. Since $A \underset{R}{\otimes} K = \Delta$. This finishes the proof. □

A more detailed study of $Br(R,\kappa)$ will follow; let us include here some results concerning the structure of κ-Azumaya algebras.

III.3.14. Remark. Let M be a two-sided A-module which is κ-closed as an R-module, then $M^{(A)}$ is κ-closed as well. Indeed, $Q_\kappa(M) = M$ entails that $Q_\kappa(M^{(A)}) \subset M$. If $m \in Q_\kappa(M^{(A)})$ then $Im \subset M^{(A)}$ for some $I \in L(\kappa)$, hence $I(ma - am) = 0$ for all $a \in A$. Since M is κ-torsion free, $ma = am$ for all $a \in A$ follows, i.e. $m \in M^{(A)}$. By local considerations one deduces (not so easily) that for a κ-Azumaya algebra A and a κ-closed (as an R-module) two-sided A-module M, we have $M \cong Q_\kappa(A \underset{R}{\otimes} M^{(A)})$.

Let $Aut_R(A)$ be the group of R-algebra automorphisms of A and let $Inn(A)$ be the subgroup of inner automorphisms of A. As before we let $R = Z(A)$ where A is a κ-Azumaya algebra over R (i.e. we assume $R = Q_\kappa(R)$).

III.3.15. THEOREM. *If A is a κ-Azumaya algebra over R then the following is an exact sequence of groups :*

$$1 \to Inn(A) \to Aut_R(A) \xrightarrow{p} Pic(R,\kappa).$$

PROOF. The proof will be a modification of the proof of the original Skolem-Noether-Rosenberg-Zelinsky theorem. If $\alpha,\beta \in Aut_R(A)$ then ${}_\alpha A_\beta$ is the $A \perp A^0$-module which is nothing but A as an additive group and multiplication induced by $(a \otimes b).x = \alpha(a)\, x\, \beta(b)$. It is clear that ${}_\alpha A_\beta$ is isomorphic to A as an R-module and in particular ${}_\alpha A_\beta$ is κ-closed. The following isomorphisms of $A \perp A^0$-modules hold :

$$Q_\kappa({}_\alpha A_1 \underset{A}{\otimes} {}_1A_\beta) = {}_\alpha A_\beta,\ {}_{\gamma\alpha}A_{\gamma\beta} = {}_\alpha A_\beta$$

for any $\gamma \in \mathrm{Aut}_R(A)$.

By the preceding remark ${}_\alpha A_1 = A \perp ({}_\alpha A_1)^{(A)}$. We write M_α for $({}_\alpha A_1)^{(A)}$ i.e. $M_\alpha = \{a \in A; \alpha(x)a = ax \text{ for all } x \text{ in } A\}$. As noted before, M_α is κ-closed and we proceed to show that M_α is κ-invertible. Pick $\alpha,\beta \in \mathrm{Aut}_R(A)$ and let $m : M_\beta \perp M_\alpha \to M_{\alpha\beta}$ be the canonical map induced by $b \perp a \mapsto ab$. An easy calculation learns that m induces an isomorphism,

$$A \perp m : A \perp M_\beta \perp M_\alpha \to A \perp M_{\alpha\beta}.$$

For each $P \in C(\kappa)$ we thus obtain an isomorphism :

$$Q_{R-P}(A \perp M_\beta \perp M_\alpha) \to Q_{R-P}(A \perp M_{\alpha\beta}),$$

i.e. :

$$Q_{R-P}(A) \underset{Q_{R-P}(R)}{\otimes} Q_{R-P}(M_\beta \perp M_\alpha) \xrightarrow{\cong} Q_{R-P}(A) \underset{Q_{R-P}(R)}{\otimes} Q_{R-P}(M_{\alpha\beta}).$$

Since $Q_{R-P}(A)$ is a free $Q_{R-P}(R)$-module we obtain isomorphisms :

$$Q_{R-P}(M_\beta \perp M_\alpha) \to Q_{R-P}(M_{\alpha\beta}).$$

The latter entails that m is an isomorphism for any $\alpha,\beta \in \mathrm{Aut}_R(A)$. In particular, taking $\beta = \alpha^{-1}$ yields that $M_\alpha \perp M_{\alpha^{-1}} = M_1 = Q_\kappa(R) = R$. Hence M_α is κ-invertible. If α is inner, say α is given by conjugation by $u \in U(A)$, then $a \in M_\alpha$ if and only if for all $x \in A$ we have : $uxu^{-1}a = ax$ i.e. $u^{-1}a \in Z(A) = R$. Consequently, if α is inner then $M_\alpha = uR \cong R$. Conversely, suppose that $f : M_\alpha \to R$ is an isomorphism, then f give rise to an isomorphism :

$$f' = A \perp f : {}_\alpha A_1 = A \perp M_\alpha \to A \perp R = R.$$

Write $f'(1) = u$, then $u\alpha(x) = f'(\alpha(x)) = f'(x.1) = xu$ i.e. $\alpha(x) = u^{-1}xu$ or α is inner. To finish the proof it is now sufficient to define p by $p(\alpha) = [M_\alpha] \in \mathrm{Pic}(R,\kappa)$.

From $M_\alpha \perp M_\beta \cong M_{\beta\alpha}$ it follows that p is a group homomorphism. Finally, since M_α is a free R-module if and only if α is inner, $\mathrm{Ker}(p) = \mathrm{Inn}(A)$ follows. □

Actually the R-algebra endomorphisms of a κ-Azumaya algebra A over R are given by :

III.3.16. PROPOSITION. *If* A *is a* κ*-Azumaya algebra over* R *then every* R*-algebra endomorphism of* A *is an automorphism.*

PROOF. Consider an R-algebra morphism $f : A \to A$. Obviously Ker f is κ-torsion free, we claim Ker f is even κ-closed. Indeed, $Q_\kappa(\text{Ker } f) \subset A$, hence if $a \in Q_\kappa(\text{Ker } f)$ then $Ia \subset \text{Ker } f$ for some $I \in L(\kappa)$. Consequently $f(Ix) = If(x) = 0$ and thus $f(x) \in \kappa(A) = 0$ implies $x \in \text{Ker } f$. Since Ker f is an ideal of A we apply Remark III.3.14. and we obtain $\text{Ker } f = Q_\kappa(A \otimes_R (\text{Ker } f)^{(A)})$. Obviously, $(\text{Ker } f)^{(A)} \subset \text{Ker } f \cap R = 0$, thus Ker f = 0. On the other hand f(A) is a κ-closed R-subalgebra of A and f(A) is a κ-Azumaya algebra with center f(R) = R. For any $P \in C(\kappa)$ we have that $Q_{R-P}(f) : Q_{R-P}(A) \to Q_{R-P}(A)$ is a $Q_{R-P}(R)$-algebra endomorphism of the Azumaya algebra $Q_{R-P}(A)$, hence $Q_{R-P}(f)$ is an automorphism. By exactness of Q_{R-P} we have $Q_{R-P}(f(A)) = Q_{R-P}(A)$. So finally $Q_\kappa(f(A)) = A$ implies $f(A) = Q_\kappa(f(A)) = A$. □

In the final part of this section we focus on splitting rings for κ-Azumaya algebras and we aim to derive a relative version of the Skolem-Noether theorem and some theory on crossed products over (κ-)Galois extensions.

Throughout this section we now assume that R is κ-closed. If S is an extension of R then we define $\bar{\kappa}$ on S-mod by taking for the class of $\bar{\kappa}$-torsion modules exactly those S-modules N which are κ-torsion when viewed as an R-module by restriction of scalars. If S is commutative then we know that $Q_\kappa(S)$ is an S-module and that $Q_\kappa(S) = Q_{\bar{\kappa}}(S)$, so in that case we only refer to $Q_\kappa(S)$. It is also clear that a prime ideal P of S will be in $K(\bar{\kappa})$ if and only if $p = P \cap R$ is in $K(\kappa)$ (recall that $K(\kappa)$ is the complement of $L(\kappa)$ in Spec R). Unless we have the lying over property for the extension $R \hookrightarrow S$ it is not necessarily true that a $P \in C(\bar{\kappa})$

intersects R in a $p = P \cap R \in C(\kappa)$. Nevertheless we have $p \in K(\kappa)$ and this is sufficient for most results; moreover usually S is a finite module or an integral extension over R and then the lying over property does hold.

III.3.17. LEMMA. *Let* $R \hookrightarrow S$ *be a* κ*-separable extension of commutative rings. Suppose that* $M \in S$*-mod is a* κ*-quasiprojective* R*-module by restriction of scalars, then* M *is a* $\bar{\kappa}$*-quasiprojective* S*-module.*

PROOF. If $P \in C(\bar{\kappa})$ then $p = P \cap R \in K(\kappa)$. Now $Q_{S-P}(M)$ is a further localization of $Q_{R-p}(M)$ and the latter is by assumption a projective $Q_{R-p}(R)$-module since $p \in K(\kappa)$. The fact that $Q_{R-p}(S)$ is a separable extension of $Q_{R-p}(R)$ entails that $Q_{R-p}(M)$ is a projective $Q_{R-p}(S)$-module and then $Q_{S-P}(M) \cong Q_{R-P}(M) \otimes_{Q_{R-p}(S)} Q_{S-P}(S)$ is clearly a projective Q_{S-} (S)-module. □

III.3.18. Remark. The existence of an M as in the lemma entails that for each $p \in C(\kappa)$ there is a $P \in C(\bar{\kappa})$ such that $P \cap R = p$.

PROOF. Pick $p \in C(\kappa)$. Since $Q_{R-P}(M)$ is a free $Q_{R-p}(R)$-module, $Q_{R-p}(p)Q_{R-p}(M) \neq Q_{R-p}(M)$. Suppose that $Q_{R-p}(p)Q_{R-p}(S) = Q_{R-p}(S)$ holds, then the fact that $Q_{R-p}(M)$ is also a $Q_{R-p}(S)$-module entails the contradiction :

$$Q_{R-p}(p)Q_{R-p}(M) = Q_{R-p}(p)Q_{R-p}(S)Q_{R-p}(M) = Q_{R-p}(M).$$

But $H = Q_{R-p}(p)Q_{R-p}(S) \neq Q_{R-p}(S)$ entails that there exists a maximal ideal of $Q_{R-p}(S)$, H' say, such that $H' \cap Q_{R-p}(R) = Q_{R-p}(p)$. Consequently, $H = j^{-1}(H')$ (where $j : S \to Q_{R-p}(S)$ is the canonical localization morphism) is an ideal of S which is prime and maximal with respect to the property $H \cap R = p$, i.e. $H \in C(\bar{\kappa})$. □

Let us now turn to commutative subrings of κ-Azumaya algebras.

III.3.19. THEOREM. *Let* A *be a* κ*-Azumaya algebra over* R *and let* S *be a* *maximal commutative subring of* A, *then* $Q_\kappa(A \otimes_S S) \cong \mathrm{End}_S(A)$ *as* S-*algebras.*

PROOF. Obviously, S is κ-torsion free, i.e. $S \subset Q_\kappa(S) \subset A = Q_\kappa(A)$. Since $Q_\kappa(S)$ is commutative $S = Q_\kappa(S)$ follows, therefore S is κ-closed. Consider the exact sequence

$$0 \to \kappa(A^e) \to A^e \xrightarrow{j} Q_\kappa(A^e) = \mathrm{End}_R(A).$$

From κ-flatness of A (and A^0) it follows that $1 \otimes S$ is embedded in A^e but then the fact that $\kappa(A^e) \cap (1 \otimes S) = 0$ yields that $1 \otimes S$ may be embedded in $Q_\kappa(A^e)$. Again by κ-flatness of A we obtain inclusions :

$$A = A \otimes_R R \hookrightarrow Q_\kappa(A \otimes_R S) \hookrightarrow Q_\kappa(A \otimes_R A^0) \cong \mathrm{End}_R(A).$$

Since Q_κ is a central localization we have :

$$Q_\kappa(A \otimes_R S) \subset Q_\kappa(A^e)^{(1 \otimes S)}.$$

With the obvious identifications, we have :

$$Q_\kappa(A^e)^{(1 \otimes S)} \cong \mathrm{End}_R(A)^{(S)} = \mathrm{End}_S(A).$$

Write $B = \mathrm{End}_S(A) \cong Q_\kappa(A^e)^{(1 \otimes S)}$; applying Remark III.3.14. to B yields $B = Q_\kappa(A \otimes B^{(A)})$. Exactness of $(-)^{(A)}$ entails : $B^{(A)} \subset Q_\kappa(A^e)^{(A)}$. If $\bar{x} \in Q_\kappa(A^e)^{(A)}$ then for some $I \in L(\kappa)$ we have $I\bar{x} \subset (A^e)^{(A)} \bmod \kappa(A^e)$ (again using exactness of $(-)^{(A)}$). But the latter ring is just $(1 \otimes A^0) \bmod (\kappa(A^E) \cap (1 \otimes A^0))$. Therefore $\bar{x}$ belongs to the image of A^0 in $Q_\kappa(A^e)$. Since $1 \otimes S \hookrightarrow Q_\kappa(A^e)$, and since $\bar{x}$ commutes with $\mathrm{Im}(1 \otimes S)$ in $Q_\kappa(A^e)$ while $\mathrm{Im}(1 \otimes S)$ is a maximal commutative subring of $\mathrm{Im}(A^0)$, it follows that $\bar{x} \in 1 \otimes S$. Consequently, $B^{(A)} \cong 1 \otimes S$ and therefore $B \cong Q_\kappa(A \otimes_R S)$. □

If M is κ-quasiprojective over R then $[M : R]_\kappa$ denotes the <u>κ-local rank function</u> $C(\kappa) \to \mathbb{N}$, given by $p \mapsto [M_p : R_p]$. It is clear that this κ-local rank function may be extended to $K(\kappa)$ and that it satisfies most properties of the usual rank function but with the domain restricted to

$C(\kappa)$. Consequently some properties of the local rank function deriving from topological conditions on Spec R, like connectedness etc..., have to be modified, but we do not use this here.

III.3.20. <u>PROPOSITION</u>. *Let* A *be a* κ-*Azumaya algebra over* R *and consider a maximal commutative subring* S *of* A. *If* A *is* $\overline{\kappa}$-*quasiprojective over* S *then* S *is* κ-*quasiprojective over* R *and* $[A : R]_\kappa = [A : S]_{\overline{\kappa}}^2$.

<u>PROOF</u>. Since A is κ-finitely generated over R it is certainly $\overline{\kappa}$-finitely generated as an S-module. From $S \subset A$ it is clear that A is also S-faithful. By assumption A is also $\overline{\kappa}$-quasiprojective in S-mod so for all $P \in C(\overline{\kappa})$ we have that $Q_{S-P}(A)$ is finitely generated in $Q_{S-P}(S)$-mod and also faithful and projective as a $Q_{S-P}(S)$-module. A well-known result (cf. [ ], 1.9., p. 76) entails that $Q_{S-P}(A)$ is faithfully projective over $Q_{S-P}(S)$ and $Q_{S-P}(S)$ is a direct summand of $Q_{S-P}(A)$. Consider :

$$A \underset{S}{\otimes} Q_{R-p}(S) = A \underset{S}{\otimes} S \underset{R}{\otimes} Q_{R-p}(R) \cong A \underset{R}{\otimes} Q_{R-p}(R).$$

This is an Azumaya algebra containing $Q_{R-p}(S)$ as a maximal commutative subring. If we write $B = \underset{P \text{ over } p}{\oplus} Q_{S-P}(S)$, the sum ranging over the prime ideals P of S such that $P \cap R = p$, then B is a faithfully flat $Q_{R-p}(S)$-module and a commutative $Q_{R-p}(S)$-algebra. Now,

$$(A \underset{S}{\otimes} Q_{R-p}(S)) \underset{Q_{R-p}(S)}{\otimes} B = \underset{P \text{ over } p}{\oplus} A \underset{S}{\otimes} Q_{S-P}(S)$$

contains $\underset{P \text{ over } p}{\oplus} Q_{R-P}(S)$ as a direct summand. By faithfully flat descent, $A \underset{S}{\otimes} Q_{R-P}(S)$ contains $Q_{R-p}(S)$ as a direct summand.
Thus $Q_{R-p}(A) \cong A \underset{S}{\otimes} Q_{R-p}(S)$ is a free $Q_{R-p}(R)$-module and $Q_{R-p}(S)$ is certainly a projective $Q_{R-p}(R)$-module. Consequently S is κ-quasiprojective in R-mod. Because A is $\overline{\kappa}$-quasiprojective over S it is easy to verify the following equalities :

$$[A : R]_\kappa = [Q_\kappa(A \underset{R}{\otimes} S) : S]_{\overline{\kappa}} = [\mathrm{End}_S(A) : S]_{\overline{\kappa}} = [A : S]_{\overline{\kappa}}^2.$$

Because S is κ-quasiprojective over R :

$$[A : R]_\kappa = [Q_\kappa(A \underset{R}{\otimes} S) : S]_{\overline{\kappa}} = [Q_\kappa(A \underset{S}{\otimes} S \underset{R}{\otimes} S) : R]_\kappa$$

$$= [A : S]_{\overline{\kappa}}.[Q_\kappa(S \underset{S}{\otimes} S) : S]_{\overline{\kappa}} = [A : S]_{\overline{\kappa}}.[S : R]_\kappa.$$

Note that in checking these equalities one uses the fact that a further localization (at a prime ideal) of some module which is already free of a certain rank remains free of the same rank. □

III.3.21. <u>COROLLARY</u>. *If* A *is a* κ*-Azumaya algebra over* R *and* S *a* κ*-separable extension of* R *which is a maximal commutative subring of* A *then* A *is* $\overline{\kappa}$*-quasiprojective over* S *and the results of the foregoing proposition hold in this situation.*

<u>PROOF</u>. If $p \in C(\kappa)$ then $Q_{R-p}(S)$ is a separable extension of $Q_{R-p}(R)$ and $Q_{R-p}(A)$ is an Azumaya algebra containing $Q_{R-p}(S)$ as a maximal commutative subring. Therefore $Q_{R-p}(A)$ is a projective $Q_{R-p}(S)$-module. For any $P \in \text{Spec } S$ lying over p (i.e. $P \in K(\overline{\kappa})$) $Q_{S-P}(A)$ is a projective $Q_{S-P}(S)$-module and therefore A is $\overline{\kappa}$-quasiprojective over S. □

III.3.22. <u>COROLLARY</u>. *If* A *is a* κ*-Azumaya algebra over the Noetherian (*κ*-closed) ring* R, *and* S *is a maximal commutative subring of* A *such that either* A *is* $\overline{\kappa}$*-quasiprojective over* S *or* S *is* κ*-separable over* R *then* S *is* κ*-quasiprojective over* R, S *and* A *are finitely generated* R*-modules and* $[A : R]_\kappa = [S : R]_\kappa^2$.

<u>PROOF</u>. An easy consequence of foregoing results, taking into account Corollary III.3.3. □

The final part of this section is concerned with the relative version of the crossed product theorem. First a useful observation :

III.3.23. <u>Remark</u>. Consider a κ-Galois extension S of R and let $P \in \text{Spec}(S)$, $p \in \text{Spec } R$, $p = P \cap R$. Then $p \in C(\kappa)$ if and only if $P \in C(\overline{\kappa})$.

PROOF. Suppose $p \in C(\kappa)$ and $P \notin C(\bar{\kappa})$ i.e. there is a $Q \in C(\bar{\kappa})$, $P \subset Q$. Clearly $p = Q \cap R$ and $Q_{R-p}(P) = Q_{R-p}(Q)$ follows from the incomparability property for prime ideals of $Q_{R-p}(S)$, the latter being a finitely generated $Q_{R-p}(R)$-module. If $q \in Q - P$ then $cq \in P$ for some $c \notin p$ implies that $q \in P$, a contradiction.
Conversely, suppose that $P \in C(\bar{\kappa})$ and $p \notin C(\kappa)$ i.e. $p \subset q$ with $q \in C(\kappa)$. Localizing at q and using the going up property for $Q_{R-q}(R) \hookrightarrow Q_{R-q}(S)$ we find Q in S such that $Q \cap R = q$ and $Q \supset P$. Therefore $Q = P$ follows, since $P \in C(\bar{\kappa})$ and $Q \notin L(\bar{\kappa})$, i.e. $q = p$. □

III.3.24. COROLLARY. *If* S *is a* κ-*Galois extension of* R *then* $\bar{\kappa} = \inf\{\kappa_{S-P}; P \in C(\bar{\kappa})\} = \inf\{\bar{\kappa}_{R-p}; p \in C(\kappa)\}$. □

If S is a κ-Galois extension of R, let U(S) be the group of units of S and let the map $c : G \times G \to U(S)$ represent a 2-cocycle for $G = \mathrm{Gal}_\kappa(S/R)$. We form the crossed product algebra $B = S[u_\tau; \tau \in G]$ as follows. Let u_τ, $\tau \in G$, be a symbol and define multiplication on the free S-module $S[u_\tau; \tau \in G]$ by putting : $u_\tau s = \tau(s)u_\tau$ for $\tau \in G$, $s \in S$ and $u_\delta u_\gamma = c(\delta.\gamma)u_{\delta\gamma}$ for $\delta, \gamma \in G$, and extending this R-linearly. Since S is κ-closed and B being a free S-module of finite rank, it follows that B is κ-closed. For each $p \in C(\kappa)$ we have that $Q_{R-p}(S)$ is a Galois extension of $Q_{R-p}(R)$ with Galois group G, cf. Proposition III.2.6. Consequently $Q_{R-p}(B)$ is a usual crossed product of G and $Q_{R-p}(S)$ i.e. an Azumaya algebra over $Q_{R-p}(R)$. Thus B is a κ-Azumaya algebra over its center R. Changing the symbol u_τ into $f_\tau\, u_\tau$ with $f_\tau \in U(S)$ obviously yields a 2-cocycle d which is equivalent to c and a crossed product algebra $B' = S[v_\tau; \tau \in G]$, with $v_\tau = f_\tau u_\tau$, which is R-algebra isomorphic to B. Conversely if we are given an R-algebra isomorphism φ,

$$\varphi : B = S[u_\tau; \tau \in G] \to B' = S[v_\tau; \tau \in G],$$

such that $\varphi|S$ is the identity on S, then the 2-cocycles c resp. d of B resp. B' are equivalent (this is just a cocycle computation formally similar to the proof of Lemma 7.5. in [68]). □

III.3.25. LEMMA. *Let* S *be a* κ*-Galois extension of* R *with* κ*-Galois group* G *and assume that every* R*-algebra automorphism of* $S[u_\tau;\tau \in G,1]$ *constructed with trivial 2-cocycle* $c(\delta,\gamma) = 1$ *for* $\delta,\gamma \in G$*, is an inner automorphism, then* $H^1(G,U(S)) = 0$.

PROOF. Check along the lines of Theorem 7.6. in [68]. □

III.3.26. LEMMA. (Relative version of the Skolem-Noether lemma)
Let A *be a* κ*-Azumaya algebra over a noetherian ring* R *containing a* κ*-Galois extension* S *of* R *with Galois group* G *as a maximal commutative subring. If* $\mathrm{Pic}(S,\overline{\kappa}) = 0$ *then every* $f \in G$ *extends to an inner automorphism of* A.

PROOF. Since R is noetherian, S and A are finitely generated R-modules by III.3.22., hence in particular S is noetherian as well. Moreover, in the definition of relative progenerators over R and S we only need relative finite generation instead of presentation. Now, from III.3.19. it follows that there are isomorphisms $A \perp S = \mathrm{End}_S(A) = S \perp A^{opp}$. Let us write B for $\mathrm{End}_S(A)$, from now on. If $\underline{A}$ is the set underlying A, then we may define an action of $A \otimes S$ on $\underline{A}$ by putting $(a \otimes s)\underline{x} = \underline{axf(x)}$ (underlined elements are in $\underline{A}$). It is clear that $\underline{A}$ is thus endowed with an $A \perp S$-module structure. Moreover, $\underline{A}$ is a left B-module, if we put $\alpha.\underline{x}^{-1} = \underline{j(\alpha)x}$, where $j : S \perp A^{opp} \tilde{\to} B$ is the canonical isomorphism. Now $\underline{A}$ is a κ-closed B-module and $\underline{A}$ is a progenerator in R-mod, so it follows that there is an isomorphism of left B-modules $\tau : A \to Q_\kappa(A \otimes_S N)$, for some κ-closed S-module N, since A is also a κ-progenerator over S. However, $\underline{A}$ is a right S-module by putting $\underline{x}.s = \underline{xf(s)}$ and if we view S as embedded in B by the identification $s \leftrightarrow (x \leftrightarrow xs)$, then $\underline{A}$ is a (B,S)-bimodule with the property that the left action of S, stemming from the embedding of S in B and the right action defined above agree. So N is a right S-module as well and τ is also an isomorphism of right S-modules. Since $\underline{A}$ is a κ-quasiprojective R-module and $A \perp S$ is κ-quasiseparable over R, it follows from local considerations that N is κ-quasiprojective as an S-module. Moreover, since both A^* and $\underline{A}$ are κ-progenerators over S, we have $N = A^* \perp_B \underline{A} = Q_\kappa(A^* \otimes_B \underline{A})$, cf. (III.1.13.), hence N is κ-finitely generated in S-mod, hence a $\bar{\kappa}$-progenerator. Consider $Q_\kappa(N \otimes_S N^*)$, then as each $P \in C(\bar{\kappa})$ lies over some $p \in C(\kappa)$, we have

$$Q_{S-P}(N \otimes_S N^*) = Q_{S-P}(N) \underset{Q_{S-P}(S)}{\otimes} Q_{S-P}(N^*) = Q_{S-P}(N) \underset{Q_{S-P}(S)}{\otimes} Q_{S-P}(N)^*.$$

It is clear that $Q_{S-P}(N)$ is a projective $Q_{S-P}(S)$-module and it will follow that $Q_{S-P}(N)$ has rank one over $Q_{S-P}(S)$ if we show that $\underline{A}$ and A have the same rank at each $P \in C(\kappa)$, since $\underline{A} = Q_\kappa(A \otimes_S N)$. But, as we have seen above

$$[A : S]_\kappa^2 = [A : S]_\kappa . [S : R]_\kappa = [A : S]_\kappa . n,$$

so $[A : S]_\kappa \equiv n$ on $C(\kappa)$. Now, $[\underline{A} : S]_\kappa = [A : S]_{f_*(\kappa)}$, meaning that $[Q_{S-P}(\underline{A}) : Q_{S-P}(S)] = [Q_{S-f(P)}(A) : Q_{S-f(P)}(S)]$ for all $P \in C(\kappa)$, so $[\underline{A} : S]_\kappa = [A : S]_{f_*(\kappa)} = [A : S]_\kappa = n$.
Consequently, $Q_{S-P}(N \otimes_S N^*) \cong Q_{S-P}(S)$ for all $P \in C(\bar{\kappa})$, hence N is a κ-invertible S-module. By assumption we thus have $N \cong S$, therefore we have a B-isomorphism $g : \underline{A} \to A \cong Q_\kappa(A \otimes_S N)$, i.e. $g(\underline{axf(s)}) = ag(x)s$ for all $s \in S$ and x,a in A. Put $g(1) = u$ and let $\underline{v}$ be such that $g(\underline{v}) = 1$, then for $\underline{a} = \underline{v}$ and $\underline{x} = \underline{s} = \underline{1}$ we obtain $1 = g(\underline{v}) = vu$ and $g(\underline{uv}) = g(\underline{1})$ follows. Therefore, $uv = 1$ and u is a unit. Finally, putting $\underline{a} = \underline{x} = \underline{1}$, we obtain $f(s) = usu^{-1}$. □

III.3.27. THEOREM. (**Relative crossed product**).
Let A *be a* κ*-Azumaya algebra over a noetherian ring* R *containing the* κ*-Galois extension* S *of* R *as a maximal commutative subring. Assume* $Pic(S,\bar{\kappa}) = 0$. *Then there exists a 2-cocycle* $c : G \times G \to U(S)$ *such that* $A \cong S[u_\tau; \tau \in G, c]$.

PROOF. By the lemma, every $\sigma \in G$ extends to an inner automorphism of A given by a unit, u_σ say, i.e. $\sigma(s) = u_\sigma s u_\sigma^{-1}$ for all $s \in S$. Choose $u_1 = 1$ and define $c(\sigma,\tau) = u_\sigma u_\tau u_{\sigma\tau}^{-1}$ for all $\sigma,\tau \in G$. Obviously $c(\sigma,\tau)$ induces the identity on S by conjugation, i.e. $c(\sigma,\tau)$ commutes with S and therefore $c(\sigma,\tau) \in U(S)$. That $c : G \times G \to U(S)$, $(\sigma,\tau) \to c(\sigma,\tau)$ represents a 2-cocycle is easily verified (by direct calculation). The crossed product R-algebra $\Delta = [v_\tau;\tau \in G,c]$ is isomorphic to A. Indeed, consider $h : \Delta \to A$ given by $s \mapsto s$ for $s \in S$, $v_\tau \mapsto u_\tau$ for $\tau \in G$. Clearly, $Q_{R-p}(h)$ is injective for all $p \in C(\kappa)$, hence h is injective because both Δ and A are κ-closed. From the Remark III.3.14. it follows that : $A \cong Q_\kappa(\Delta \underset{R}{\otimes} A^{(h(\Delta))})$. But $A^{(h(\Delta))} \subset A^{(S)} \subset S$, thus $A^{(h(\Delta))} \subset S^{h(\Delta)} \subset R$, the latter inclusion following from the fact that $u_\tau \in h(\Delta)$ for all $\tau \in G$. So finally $A \cong Q_\kappa(\Delta) = \Delta$ follows, as claimed.

It may be expected that this relative crossed product theorem can be applied in the following situations :

1° Sheaves of Azumaya algebras which are well-behaved on some open set of the base space,

2° Azumaya algebras over Krull domains which may be split by commutative extensions having the "Galois property" for prime ideals of height one.

III.4. Some Exact Sequences.

The result in this section relate to those of [67]. Throughout R will be a Noetherian ring and $\tau \leq \kappa$ are idempotent kernel functors on R-mod. We write $P(\tau)$, $P(\kappa)$ for the category of τ-progenerators, resp. κ-progenerators. Localization at κ defines a functor $Q_\kappa : P(\tau) \to P(\kappa)$; it is also clear that any κ-closed module is also τ-closed.

III.4.1. PROPOSITION. *Consider* $E,F \in P(\kappa)$ *such that there exist* $E_1,G \in P(\tau)$ *with the property that* $E = Q_\kappa(E_1)$ *and* $Q_\kappa(E \underset{R}{\otimes} F) = Q_\kappa(G)$, *then* $F \in P(\tau)$.

PROOF. Clearly F is τ-closed. Consider a τ-open G_1 in G which is finitely generated and write $\overline{G} = G/\kappa(G)$, $\overline{G}_1 = G_1/\kappa(G_1)$. Then $\overline{G}_1$ is τ-open in $\overline{G}$ and $\overline{G}_1$ is finitely generated. We have :

$$\overline{G} \subset Q_\kappa(G) = Q_\kappa(E \underset{R}{\otimes} F) = Q_\kappa(E_1 \underset{R}{\otimes} F),$$

while $E_1 \underset{R}{\otimes} F$ is κ-torsion free since it is κ-flat. Put $H = (E_1 \underset{R}{\otimes} F) \cap \overline{G}_1$. Then H is finitely generated since R is Noetherian and H is τ-open in $E_1 \underset{R}{\otimes} F$. Assume H is generated by a finite family $\{e_i \otimes f_i\}$, $e_i \in E_1$, $f_i \in F$, and write $F' = \sum_i Rf_i$. From τ-flatness of E_1 it follows that the canonical $\varphi : E_1 \underset{R}{\otimes} F' \to E_1 \underset{R}{\otimes} F$ has τ-torsion kernel. By definition of F' we have : $Im(\varphi) \subset H \subset E_1 \underset{R}{\otimes} F$. The inclusion $F' \subset F$ induces an isomorphism : $Q_\tau(E_1 \underset{R}{\otimes} F') = Q_\tau(E_1 \otimes F)$. For each $P \in \mathcal{C}(\tau)$ we know that $Q_{R-P}(E_1)$ is a free $Q_{R-P}(R)$-module and consequently the map $Q_{R-P}(F') \to Q_{R-P}(F)$ is an isomorphism. Therefore $Q_\tau(F') = Q_\tau(F)$ i.e. F' is τ-open in F or F is τ-finitely generated. Because R is Noetherian F is also τ-finitely presented. The statement of the proposition is now easily verified. □

Let Γ denote the set of isomorphism classes of objects E in $\mathcal{P}(\kappa)$ such that $End_R(E)$ is a τ-Azumaya algebra. Define a relation ~ on Γ as follows : $E \sim E_1$ if and only if there are P,Q in $\mathcal{P}(\tau)$ such that $Q_\kappa(E \underset{R}{\otimes} P) \cong Q_\kappa(E_1 \underset{R}{\otimes} Q)$. This is obviously an equivalence relation. The product $\underset{\kappa}{\perp}$ defines an abelian group structure on $\Gamma/\sim$. The inverse of a class [E] is [E*]. Indeed $E \underset{\kappa}{\perp} E^* = End_R(E)$ and $End_R(E) \in \mathcal{P}(\tau)$ by assumption, so we may apply the foregoing proposition. We write $Bcl(R,\kappa,\tau)$ for $\Gamma/\sim$. There is a group morphism $j : Pic(R,\kappa) \to Bcl(R,\kappa,\tau)$, $P \mapsto [P]$. Indeed, if $P \in \underline{Pic}(R,\kappa)$ then $End_R(P) = Q_\kappa(R) = R$ and obviously R is a τ-Azumaya algebra over R. On the other hand there is a group morphism

$$\alpha : Bcl(R,\kappa,\tau) \to Br(R,\tau),\ [E] \to [End_R(E)].$$

Moreover, $Q_\kappa : \mathcal{P}(\tau) \to \mathcal{P}(\kappa)$ induces group morphisms : $i : Pic(R,\tau) \to Pic(R,\kappa)$ and $\beta : Br(R,\tau) \to Br(R,\kappa)$. These morphisms may be related by the following :

III.4.2. THEOREM. *With notations as above, we obtain the following exact sequence of groups :*

$$Pic(R,\tau) \xrightarrow{i} Pic(R,\kappa) \xrightarrow{j} Bcl(R,\kappa,\tau) \xrightarrow{\alpha} Br(R,\tau) \xrightarrow{\beta} Br(R,\kappa).$$

PROOF. Along the lines of a similar statement in [67], let us present a sketch of the proof here. That the above sequence is a complex is evident. In order to check exactness at $Pic(R,\kappa)$ assume that $P \in \underline{Pic}(R,\kappa)$ is such that $j([P]) = 0$. Then there exist τ-progenerators E,F such that $Q_\kappa(P \otimes Q_\kappa(E)) = Q_\kappa(F)$. That P is a τ-progenerator follows from Proposition III.4.1. But then $Q_\tau(P^* \underset{R}{\otimes} P) = End_R(P)$ and $End_R(P) = Q_\kappa(R) = R$. It follows that P is τ-invertible.

To check exactness at $Bcl(R,\kappa,\tau)$, consider $E \in \Gamma$ and assume that $\alpha([E]) = 0$. Then $End_R(E) \sim R$ in $Br(R,\tau)$. We may find $P \in \mathcal{P}(\tau)$ such that $End_R(E) = End_R(P)$. Because E is a κ-progenerator, $End_R(E)$ is κ-closed, hence $End_R(P) = Q_\kappa(End_R(P)) = End_R(Q_\kappa(P))$ where $Q_\kappa(P)$ is now a κ-progenerator. Put $B = End_R(Q_\kappa(P))$, then E is a B-module in the obvious way. The properties of $\underset{\kappa}{\perp}$ imply :

$$Hom_B(Q_\kappa(P),E) \perp Q_\kappa(P) = (Q_\kappa(P)^* \perp E) \perp Q_\kappa(P) =$$

$$(E \underset{B^0}{\perp} Q_\kappa(P)^*) \perp Q_\kappa(P) = E \underset{B^0}{\perp} End_R(Q_\kappa(P)) = B \underset{B}{\perp} E = E$$

and $F = Hom_B(Q_\kappa(P),E) \in \mathcal{P}(\kappa)$.

But, $F \perp Q_\kappa(P) = E$ and $End_R(Q_\kappa(P)) = End_R(E)$ imply that $End_R(F) = R$, i.e. $F \in \underline{Pic}(R,\kappa)$. The obvious fact $[F] = [E]$ then entails the exactness of the sequence at $Bcl(R,\kappa,\tau)$.

Finally, if $[A] \in Ker\ \beta$ then $Q_\kappa(A) = End_R(E)$ for some κ-progenerator E but then $[A] \in Im(\alpha)$, and this finishes the proof. □

By taking κ,τ to be special kernel functors one may derive from the sequence in Theorem III.4.2. many interesting corollaries, e.g. τ trivial, κ arbitrary or τ arbitrary, κ trivial, or $\tau = \inf\{\kappa_{R-P}; P \in X^1(R)\}$ $\tau \leq \kappa$ etc... We leave verification of all variations to the reader and we will return to some special case when needed in the sequel.

CHAPTER IV
APPLICATION TO SHEAVES, RINGED SPACES AND SCHEMES

IV.1. Brauer Groups of Ringed Spaces.

In this section we recall some notions and techniques stemming from work of A. Grothendieck [47, I, II, III], and B. Auslander [8]. Some elementary theory of sheaves may be found in [42], [79].

Let X be a topological space together with a sheaf of commutative rings $\mathcal{O}_X$. The category of sheaves of $\mathcal{O}_X$-modules is denoted by $\mathcal{O}_X$-Mod. It is a tradition to denote sheaf theoretical equivalents of ring theoretical objects by the same word but written with a capital, i.e. Module will mean sheaf of modules etc...

A sheaf of $\mathcal{O}_X$-modules, M say, is locally projective of finite type if it is locally a direct summand of a sheaf of free $\mathcal{O}_X$-modules of finite rank, i.e. for each $x \in X$, there exists an open neighborhood U of x together with sheaf morphisms $\psi : M|U \to \mathcal{O}_X^n|U$, $\varphi : \mathcal{O}_X^n|U \to M|U$ for some $n \in \mathbb{N}$, such that $\varphi \circ \psi$ coincides with the identity on $M|U$. By $M|U$ we mean the restriction of the sheaf M to the open set U, cf. [42], [79]. The pair $(X,\mathcal{O}_X)$ is called a ringed space. A particular case is obtained by taking for X the spectrum Spec(R) of a commutative ring with its Zariski topology, while $\mathcal{O}_X$ is the usual structure sheaf of R. If $(X,\mathcal{O}_X)$ is of this special form then we call this ringed space an affine ringed space. The latter may be used in constructing the more general schemes in the sense of A. Grothendieck, i.e. a scheme will be a ringed space which is "locally" an affine ringed space. Schemes are examples of <u>locally ringed spaces</u> because for each $x \in X$ the stalk $\mathcal{O}_{X,x}$ of $\mathcal{O}_X$ at x is a local commutative ring. The set of sections of a sheaf S over an open set U will be denoted by $S(U)$, the stalk of S at $x \in X$ will be denoted by S_X, where $S_X = \varinjlim_{x \in U} S(U)$.

IV.1.1. <u>PROPOSITION</u>. *Let* $(X, \mathcal{O}_X)$ *be a ringed space and* $M \in \mathcal{O}_X$*-Mod. The following statements are equivalent :*

1. M *is locally projective of finite type.*
2. M *is locally of finite type and for each open subset* U *of* X*, the functor* $\mathrm{Hom}_{\mathcal{O}_X|U}(M|U,-)$ *is exact in* $\mathcal{O}_X|U$*-Mod.*
3. M *is finitely presented and for all* $x \in X$*,* M_x *is a projective* $\mathcal{O}_{X,x}$*-module.* □

IV.1.2. <u>LEMMA</u>. *Let* $X = \mathrm{Spec}(R)$ *for some commutative ring* R *and let* $\mathcal{O}_X = \mathcal{O}_R = \widetilde{R}$ *be the structure sheaf of* R*. For* $M \in \widetilde{R}$*-Mod, the following statements are equivalent :*

1. M *is locally projective of finite type.*
2. M *is locally free of finite rank.* □

The statements in Lemma IV.1.2. are still equivalent if $(X, \mathcal{O}_X)$ is an arbitrary scheme because the proof may in this case be reduced to the affine case, stated above.

Observe that, if M and N are locally projective sheaves of $\mathcal{O}_X$-modules, then $Hom_{\mathcal{O}_X}(M,N)$ and also $M \otimes_{\mathcal{O}_X} N$ have that property. Moreover it is easy to derive the following canonical isomorphisms :

$$Hom_{\mathcal{O}_X}(M,M) \otimes_{\mathcal{O}_X} Hom_{\mathcal{O}_X}(N,N) \cong Hom_{\mathcal{O}_X}(M \otimes_{\mathcal{O}_X} N,\ M \otimes_{\mathcal{O}_X} N)$$

$$Hom_{\mathcal{O}_X}(M,\mathcal{O}_X) \otimes_{\mathcal{O}_X} M \cong Hom_{\mathcal{O}_X}(M,M).$$

It is evident that there is a canonical mapping $\omega : \mathcal{O}_X \to Hom_{\mathcal{O}_X}(M,M)$, defined "locally" as multiplication by $r(U) \in \underline{\mathcal{O}}_X(U)$ on the open set U of X.

The kernel of ω, denoted by $Ann_{O_X}(M)$ is called the <u>annihilator sheaf</u> of M.

If $Ann_{O_X}(M)$ vanishes then M is said to be a <u>faithful sheaf</u> of O_X-modules. In case M is finitely presented it follows that M is faithful if an only if M_x is faithful as an $O_{X,x}$-module for all $x \in X$. The sheaf $Hom_{O_X}(M,M)$ will be written $E(M)$ and by local argumentation it is clear that M is also an $E(M)$-Module. If M is a faithful locally projective O_X-Module then it is also faithful and locally projective as an $E(M)$-Module. An O_X-<u>Algebra</u> is just a sheaf of algebras over the ringed space (X,O_X). To an O_X-Algebra L there may be associated another O_X-Algebra, called the <u>opposite Algebra</u>, denoted by L^0, which is defined by $L^0(U) = (L(U))^0$ for every open set U of X. The <u>enveloping Algebra</u> of L is $L^e = L \otimes_{O_X} L^0$. Obviously, L is a left L^e-Module, the action being given locally on open sets U of X by the action of $(L(U))^e$ on $L(U)$. That this has some advantages may be clear from the following observation. If L is an O_X-Algebra which has finite type as an O_X-Module then L is finitely presented in L^e-Mod. There is a commutative diagram :

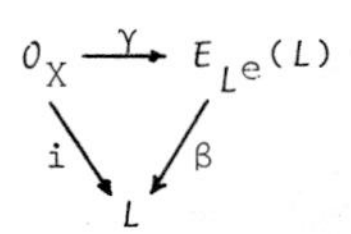

where i is the structure morphism for L, γ is defined as ω before, β is given by evaluating at the identity of L. All these morphisms are in fact monomorphisms, so we may identify $E_{L^e}(L)$ and its image $Z(L)$ in L. The latter sheaf is called the <u>center</u> of L. If γ is moreover an isomorphism then we say that L is a <u>central sheaf of O_X-algebras</u>. Clearly, a finitely presented L in O_X-Mod is central over O_X if and only if L_x is a central $O_{X,x}$-algebra for all $x \in X$.

Let us define the morphism $\eta : L^e \to Hom_{O_X}(L,L)$ locally as follows : $\eta(a \otimes b)(x) = axb$, for $x \in L(U)$, $a \otimes b \in L^e(U)$, U open in X. A central O_X-Algebra L such that L is locally projective and finitely generated as an O_X-Module is said to be <u>locally separable</u> if η is an isomorphism. For example : if M is a faithful locally projective $\mathcal{O}_X$-Module of finite type, then $Hom_{\mathcal{O}_X}(M,M)$ is locally separable.

If A and B are locally separable $\underline{\mathcal{O}}_X$-Algebras then also $A \underset{\mathcal{O}_X}{\otimes} B$ is locally separable. Moreover, if L is locally separable over $\mathcal{O}_X$ and $\mathcal{O}'_X$ is any commutative $\underline{\mathcal{O}}_X$-Algebra then $L \underset{\mathcal{O}_X}{\otimes} \mathcal{O}'_X$ is any locally separable $\mathcal{O}'_X$-Algebra.

Now consider the situation where $(X,\mathcal{O}_X)$ is a scheme. An $\mathcal{O}_X$-Algebra L is now called an <u>Azumaya algebra over X</u> if L is a coherent $\mathcal{O}_X$-Module such that for all closed points $x \in X$ we have that L_x is an Azumaya algebra over $\mathcal{O}_{X,x}$.

IV.1.3. <u>PROPOSITION</u>. *Consider a scheme $(X,\mathcal{O}_X)$ and let L be an $\mathcal{O}_X$-Algebra which is finitely generated as an $\mathcal{O}_X$-Module. The following statements are equivalent :*

1. *L is an Azumaya algebra over X*
2. *L is locally free in $\mathcal{O}_X$-Mod and for all $x \in X$, $L(x) = L_x \underset{\mathcal{O}_{X,x}}{\otimes} \mathbb{k}(x)$ is a central simple $\mathbb{k}(x)$-algebra, where $\mathbb{k}(x)$ denotes the residue field of the local ring $\mathcal{O}_{X,x}$.*
3. *L is a locally free $\mathcal{O}_X$-Module and the morphism η is an isomorphism.*
4. *There exists an étale covering $(U_i \to X)_i$ for X with the property that for each i there is an $n_i \in \mathbb{N}$ such that $L \underset{\mathcal{O}_X}{\otimes} \mathcal{O}_{U_i} = M_{n_i}(\mathcal{O}_{U_i})$.*

The proof of the foregoing proposition consists of standard techniques in sheaf theory and elementary properties of Azumaya algebras as in Chapter I.3. Note that 4. derives from the fact that Azumaya algebras always have étale splitting rings, cf. [47]. The foregoing definitions and preliminary results indicates that the definition of the Brauer group of a ring can be extended without much problems to the definition of the Brauer group of a ringed space or scheme. Indeed, let $(X,\mathcal{O}_X)$ be an arbitrary ringed space. Consider the category $C_1(X,\mathcal{O}_X)$ of locally separable $\mathcal{O}_X$-Algebras with $\mathcal{O}_X$-Algebra morphisms. Let $C'_1(X,\mathcal{O}_X)$ be the

full subcategory of $C_1(X,\mathcal{O}_X)$ consisting of objects which are sheaf isomorphic to $\mathcal{H}om_{\mathcal{O}_X}(M,M)$ for some faithful locally projective $\mathcal{O}_X$-Module M which is finitely generated. We have noted before that both $C_1(X,\mathcal{O}_X)$ and $C_1'(X,\mathcal{O}_X)$ are closed under taking tensorproducts over $\mathcal{O}_X$. We introduce an equivalence relation on $C_1(X,\mathcal{O}_X)$ as follows : if $L_1, L_2 \in C_1(X,\mathcal{O}_X)$ then L_1 is said to be equivalent to L_2 if and only if there exist E_1 and E_2 in $C_1'(X,\mathcal{O}_X)$ such that : $L_1 \underset{\mathcal{O}_X}{\otimes} E_1 \cong L_2 \underset{\mathcal{O}_X}{\otimes} E_2$.
The set of equivalence classes for the relation just defined is a group with respect to the multiplication induced by the tensorproduct over $\mathcal{O}_X$; so it is an abelian group denoted by $\mathrm{Br}(X,\mathcal{O}_X)$, or more simply $\mathrm{Br}(X)$ if the sheaf is the obvious one, and this group is called the Brauer group of the ringed space $(X,\mathcal{O}_X)$, or of X is no ambiguity is possible.
In case $(X,\mathcal{O}_X)$ is a scheme then this Brauer group is exactly the same as the group defined by A. Grothendieck in [47], because of Proposition IV.1.3.,4. Let $[L]$ denote the class (sometimes called the Brauer class) of L. As observed before, we see that $[L]^{-1} = [L^0]$ and $[\mathcal{O}_X]$ is the unit element in $\mathrm{Br}(X,\mathcal{O}_X)$.

If S is any sheaf over X then the elements of $S(U)$ are called sections over U of S, elements of $S(X)$ are called global sections of S. The fact that for every commutative ring R we have $R = \tilde{R}(X)$, i.e. R is the ring of global sections of the structure sheaf $\tilde{R}$ over Spec(R), is the basis for the fundamental duality between geometrical and algebraic aspects of algebraic geometry.
Concerning the Brauer group this duality is still valid in the affine case, due to the following result, cf. [8].

IV.1.4. PROPOSITION. *Let* X = Spec(R) *be an affine scheme, then* Br(X) = Br(R). □

The proof is rather straightforward and it is essentially obtained by showing that taking global sections of the Algebras L in Br(X) defines an isomorphism between Br(X) and $\mathrm{Br}(\tilde{R}(X)) = \mathrm{Br}(R)$. As a first application of the relative theory developed in Chapter III we are now able to describe Br(X(I)) where X(I) is a not necessarily affine open subset of X.

IV.2. Picard and Brauer Groups of Quasi-Affine Schemes.

Throughout this section R is a commutative ring, X = Spec(R) is endowed with the Zariski topology with open sets $X(I) = \{P \in X;\ I \not\subset P\}$ associated to ideals I of R. A basis for this topology is given by $\{X_f; f \in R\}$ where $X_f = X((f))$. The structure sheaf $\mathcal{O}_R$ may be defined by associating to X_f the ring R_f which is the localization of R at the multiplicative set $\{1,f,f^2,\ldots\}$. Note that it suffices to describe the sheaf on a basis of open sets, cf. [49]. To any finitely generated ideal I of R we associate the idempotent kernel functor κ_I with filter $\mathcal{L}(I) = \{L \text{ ideal of } R;\ \mathrm{rad}(L) \supset I\}$. The ring of sections of $\mathcal{O}_R$ over X(I) will be denoted by $\Gamma(X(I),\mathcal{O}_R)$. Then, for each finitely generated ideal I of R we have that $\Gamma(X(I),\mathcal{O}_R) = Q_I(R)$ where Q_I is the localization functor corresponding to the kernel functor κ_I. In a similar way we associate a structure sheaf $\tilde{M} = \mathcal{O}_M$ to each R-module M.

Let $\mathcal{S}(R)$, resp. $\mathcal{P}(R)$, be the category of sheaves, resp. presheaves, of abelian groups over X and let $i : \mathcal{S}(R) \to \mathcal{P}(R)$ be the canonical inclusion. Since i is left exact we can consider the derived functors $\mathcal{H}^q = R^q i$. For each open set U in X, taking "sections over U" is exact in $\mathcal{P}(R)$, therefore we obtain $\mathcal{H}^q(G) = H^q(U,G)$ for each sheaf of abelian groups $G \in \mathcal{S}(R)$. The presheaf $\mathcal{H}^q(G)$ is the presheaf defined by associating to each open set U of X the group $H^q(U,G)$, together with the obvious restriction morphisms.

In particular, put q = 1 and take for G the sheaf of (multiplicative) groups $\mathcal{O}_R^*$ consisting of the invertible sections of $\mathcal{O}_R$, i.e. by definition $\Gamma(U,\mathcal{O}_R^*) = \Gamma(U,\mathcal{O}_R)^*$. Let us refer to [79] for some basic facts on cohomology with values in sheaves. In this particular situation we obtain :

$$\mathcal{H}^1(\mathcal{O}_R^*)(U) = H^1(U,\mathcal{O}_R^*) = H^1(U,(\mathcal{O}_R|U)^*) = \mathrm{Pic}(U,\mathcal{O}_R|U)$$

where the latter is the Picard group of the quasi-affine scheme U, viewed as an open subscheme of $(X,\mathcal{O}_R)$, basis open sets in X the above groups are readily calculated. More generally, if R is a Noetherian ring then an open set X(I) is exactly then affine if X_I is a perfect localization i.e. a T-functor, and in this case, $\text{Pic}(X(I),\mathcal{O}_R|X(I)) = \text{Pic}(Q_I(R))$.

Elementary theory about coherent sheaves may be found in [42] or [79]. We need the following

IV.2.1. LEMMA. *Let* R *be a Noetherian ring and consider a Zariski open set* U = X(I) *of* X. *Let* M *be a coherent sheaf of groups over* U, *then :*
1. *There is a coherent sheaf* M' *over* X *such that* $M'|U = M$.
2. *If* M' *and* M" *are R-modules such that* $M'|U = M''|U$ *then* $Q_I(M') = Q_I(M'')$.

PROOF. 1. If $j : U \to X$ is the canonical inclusion then it suffices to verify that i_*M is quasicoherent on X. Indeed a quasicoherent sheaf over a Noetherian scheme is the direct union of its coherent subsheaves and one of these will satisfy the requirements of 1., cf. [79] for details.
2. For any X(I) we have $\Gamma(X(I),\widetilde{M}) = Q_I(M)$; therefore the proof of 2. is easy. □

IV.2.2. Remark. We shall return to a similar problem of extending coherent and quasicoherent sheaves in a subsequent section; there U will not be affine or not even open and so the techniques involved have to be modified.

Let us define a presheaf $\mathcal{Pic}_R$ on X by associating to each open set X(I) of X the group of sections $\Gamma(X(I),\mathcal{Pic}_R) = \text{Pic}(R,\kappa_I)$ i.e. the relative Picard group with respect to κ_I. If $X(J) \subset X(I) \subset X$ then the restriction morphism $\varphi_{I,J} : \text{Pic}(R,\kappa_I) \to \text{Pic}(R,\kappa_J)$ is defined by $[P] \to [Q_J(P)]$. From hereon we assume that R is a Noetherian ring, for simplicity's sake. In this case each open set X(I) of X is quasicompact. For an arbitrary basic open set X_f we have :

$$\Gamma(X_f, \mathcal{P}ic_R) = \mathrm{Pic}(R,\kappa_f) = \mathrm{Pic}(Q_f(R)) = \mathrm{Pic}(X_f, \mathcal{O}_{Q_f(R)})$$

$$= H^1(X_f, \mathcal{O}_R^*) = \Gamma(X_f, H^1(\mathcal{O}_R^*))$$

i.e. the presheaves $H^1(\mathcal{O}_R^*)$ and $\mathcal{P}ic_R$ coincide on a basis of open sets of X. In order to derive the following result we need some more work because the presheaves mentioned are not even separated !

IV.2.3. THEOREM. *The presheaves of abelian groups $H^1(\mathcal{O}_R^*)$ and $\mathcal{P}ic_R$ are isomorphic.*

PROOF. We have to check whether for each ideal I of R, $\mathcal{P}ic_R(X(I)) = \mathrm{Pic}(R,\kappa_I)$, and $H^1(\mathcal{O}_R^*)(X(I)) = \mathrm{Pic}(X(I),\mathcal{O}_R^*)$.
We use the following notation : $\mathcal{O}_{R,I} = \mathcal{O}_R|X(I)$, $i : X(I) \hookrightarrow X$ is the canonical inclusion. Consider $[\mathcal{P}] \in \mathrm{Pic}(X(I),\mathcal{O}_R^*)$, then there exists $[\mathcal{Q}] \in \mathrm{Pic}(X(I),\mathcal{O}_R^*)$ such that $\mathcal{P} \underset{\mathcal{O}_{R,I}}{\otimes} \mathcal{Q} \cong \mathcal{O}_{R,I}$ and by the lemma we may find R-modules (the module structure is inherited by the (pre)sheaves constructed there !) P and Q such that $\mathcal{P} = \widetilde{P}|X(I)$, $\mathcal{Q} = \widetilde{Q}|X(I)$. For any $X_f \subset X(I)$ we have :

$$\Gamma(X_f, (P \underset{R}{\otimes} Q)^\sim) = Q_f(P) \underset{R_f}{\otimes} Q_f(Q) = \Gamma(X_f,\widetilde{P}) \underset{\Gamma(X_f,\mathcal{O}_R)}{\otimes} \Gamma(X_f,\widetilde{Q})$$

$$= \Gamma(X_f,\widetilde{P}|X(I)) \underset{\Gamma(X_f,\mathcal{O}_{R,I})}{\otimes} \Gamma(X_f,\widetilde{Q}|X(I))$$

$$= \Gamma(X_f,\mathcal{P}) \underset{\Gamma(X_f,\mathcal{O}_{R,I})}{\otimes} \Gamma(X_f,\mathcal{Q}).$$

So $(P \underset{R}{\otimes} Q)^\sim$ and $\mathcal{P} \underset{\mathcal{O}_{R,I}}{\otimes} \mathcal{Q}$, coincide on a basis of open subsets of X(I), consequently : $(P \underset{R}{\otimes} Q)^\sim|X(I) = \mathcal{P} \underset{\mathcal{O}_{R,I}}{\otimes} \mathcal{Q}$. But then we obtain :

$$Q_I(R) = \Gamma(X(I),\mathcal{O}_{R,I}) \cong \Gamma(X(I),\mathcal{P} \underset{\mathcal{O}_{R,I}}{\otimes} \mathcal{Q}) = \Gamma(X(I),(P \underset{R}{\otimes} Q)^\sim) = Q_I(P \underset{R}{\otimes} Q),$$

i.e. P and hence $Q_I(P)$ is κ_I-invertible. Moreover $Q_I(P)$ is determined by $\mathcal{P}$ because $Q_I(P) = \Gamma(X(I),\widetilde{P}) = \Gamma(X(I),\mathcal{P})$. Any monomorphism of R-modules $\nu : M' \to M$ gives rise to a sheaf morphism $\widetilde{\nu} : \widetilde{M}' \to \widetilde{M}$ which in turn determines a morphism :

$$\tilde{P} \otimes \tilde{\nu} : \tilde{P} \underset{\mathcal{O}_R}{\otimes} \tilde{M}' \to \tilde{P} \underset{\mathcal{O}_R}{\otimes} \tilde{M}$$

which is injective on $X(I)$, because for each prime ideal $p \in X(I)$ the stalk map $(\tilde{P} \otimes \tilde{\nu})_p$ is injective. Indeed, for any R-module M we see that $(\tilde{P} \underset{\mathcal{O}_R}{\otimes} \tilde{M})_p = Q_p(P) \underset{R_p}{\otimes} Q_p(M)$, where $Q_p(P)$ is a free $Q_p(R)$-module of rank one since $\tilde{P}|X(I) = \mathcal{P}$ is invertible. The morphism $\Gamma(X(I),\tilde{P} \otimes \tilde{\nu})$ is therefore injective too and this entails that :

$$\mathrm{Ker}(Q_I(P \otimes \nu)) = \mathrm{Ker}(\Gamma(X(I),(P \otimes \nu)^{\sim}) = \mathrm{Ker}(\Gamma(X(I),\tilde{P} \otimes \tilde{\nu}) = 0.$$

Left exactness of Q_I yields that $\mathrm{Ker}(P \otimes \nu)$ is κ_I-torsion and so each invertible sheaf $\mathcal{P}$ over $(X(I),\mathcal{O}_{R,I})$ is up to isomorphism associated to a unique κ_I-invertible κ_I-closed R-module $Q_I(P)$. This defines the injective map

$$\varphi : \mathrm{Pic}(X(I),\mathcal{O}_R^*) \to \mathrm{Pic}(R,\kappa_I).$$

That φ is a group morphism is clear. Take a κ_I-invertible P with $[P] \in \mathrm{Pic}(R,\kappa_I)$. Since $Q_f(P)$ is κ_f-invertible for each affine subset $X_f \subset X(I)$, each $Q_f(P)$ is an invertible $Q_f(R)$-module. To each $Q_f(P)$ we associate a quasicoherent sheaf of $\mathcal{O}_{Q_f(R)}$-modules $(Q_f(P))^{\sim}$ over X_f and these glue together well, yielding a sheaf $\mathcal{P}$ over $X(I)$ which is completely determined by P. The properties of P entail that there is a κ_I-invertible module Q such that : $Q_I(P \underset{R}{\otimes} Q) \cong Q_I(R)$. Consequently for each X_f contained in $X(I)$ we obtain :

$$Q_f(R) = Q_f(Q_I(R)) = Q_f[Q_I(P \underset{R}{\otimes} Q)] \cong Q_f(P) \underset{R}{\otimes} Q_f(Q).$$

It follows that the sheaf $\mathcal{Q}$ determined by the Q_f over X_f yields an inverse for $\mathcal{P}$. Evidently this establishes surjectivity of the map Φ. □

IV.2.4. COROLLARY. *If* R *is a Noetherian ring and* I *is an ideal of* R *then* $\mathrm{Pic}(X(I)) = \mathrm{Pic}(R,\kappa_I)$.

The second part of this section is devoted to the study of the Brauer group of a quasi-affine scheme. For notations and definitions of some basic concepts we refer to Section IV.1.

IV.2.5. PROPOSITION. *There is a bijective correspondence between the following classes :*

1. *The class of locally projective sheaves of finite type over* U = X(I).
2. *The class of* κ_I*-quasiprojective,* κ_I*-finitely generated* κ_I*-closed* R-*modules.*

PROOF. Any locally projective sheaf M over U of finite type M is coherent. Hence M is of the form $\tilde{M}|U$ for some finitely generated R-module M because of Lemma IV.2.1.
Moreover, $M = (Q_I(M))^{\sim}|U$, so we may associate $Q_I(M)$ to M. This is a well-defined correspondence and furthermore $Q_I(M)$ is κ_I-closed κ_I-finitely presented because R is assumed to be Noetherian. Since M is locally projective of finite type it follows that for each $p \in X(I)$ we have that $M_p = Q_p(Q_I(M)) = Q_p(M)$ is a finitely generated projective R_p-module. Now observe that the foregoing certainly holds for all $p \in C(\kappa_I)$. Conversely, applying some classical properties of localization it follows immediately that each object in 2. may be considered as the object of sections over X(I) obtained from a sheaf as in 1. □

IV.2.6. COROLLARY. *There is bijective correspondence between the following classes :*

1. *Locally separable central sheaves over* U = X(I).
2. κ_I*-Azumaya algebras over* R.

PROOF. For a locally separable central sheaf L over U we calculate that $Q_I(\Gamma(U,L)) = L$ is κ_I-closed, κ_I-quasiprojective and κ_I-finitely presented and since for each $p \in C(\kappa_I) \subset X(I)$ we have that $L_p = L_p$ is an Azumaya algebra over R_p, the statement is a consequence of the theory of Section III.3. □

IV.2.7. COROLLARY. *There is bijective correspondence between the following classes :*

1. *The faithful locally projective sheaves of finite type over* U = X(I).
2. *The* κ_I*-progenerators of* R*-mod.*

PROOF. Sections III.1., III.3. □

Combining the above results yields :

IV.2.8. THEOREM. *Let* $Br(X(I))$ *be the Brauer group of the quasi-affine scheme* $(X(I),\mathcal{O}_R|X(I))$ *for some ideal* I *of the Noetherian ring* R, *then* $Br(X(I)) = Br(R,\kappa_I)$.

IV.2.9. Example. Assume that R is a Noetherian ring of Krull dimension one and let $X = Spec(R)$, let X_{reg} denote the open subscheme consisting of the regular points of X, then application of the foregoing theorem yields : $Br(X_{reg}) = \bigcap_{P\in X_{reg}} Br(R_P)$. (Apply Proposition III.3.13.). For suitable choices of R and I this result may be sharpened as we shall establish in the sequel. We need some preliminary results.

IV.2.10. LEMMA. *If* R *is* κ*-closed, then any finitely generated projective* R*-module is* κ*-closed too.*

PROOF. If P is a finitely generated projective R-module then $P \oplus Q = R^n$ for some $n \in \mathbb{N}$ and $Q \in$ R-mod. Since R^n is κ-torsion free, P is κ-torsion free, and the same holds for Q. Since R^n is κ-closed we have $Q_\kappa(P) \subset R^n$. Take $y \in Q_\kappa(P) - P$. Then $Iy \subset P$ for some $I \in \mathcal{L}(\kappa)$. Writing $y = y_P + y_Q$ with $y_P \in P$, $y_Q \in Q$ then $Iy \subset P$ yields $Iy_Q = 0$ i.e. $y_Q = 0$. So we have $Q_\kappa(P) = P$ and P is κ-closed. □

IV.2.11. LEMMA. *If* R *and* κ *are such that* $Q_\kappa(R)$ *is a Noetherian domain of global dimension at most two, then any* κ*-finitely generated* κ*-closed reflexive* $Q_\kappa(R)$*-module* P *is a projective* $Q_\kappa(R)$*-module.*

PROOF. This proof follows the lines of the proof (due to L. Roberts) of a similar global result. First note that if P is reflexive then it is certainly torsion free. Since P is κ-finitely generated and R is Noetherian, it follows (see Section III.1.) that P^* is finitely generated as a $Q_\kappa(R)$-module. Write $S = Q_\kappa(R)$; consider an exact sequence in S-mod : $0 \to L \to S^n \to P^* \to 0$, for some $n \in \mathbb{N}$. Dualization of this sequence yields :

$$0 \to P^{**} = P \to S^n \to Q \to 0,$$

where Q is a submodule of L^*, i.e. Q is κ-torsion free and finitely generated as an image of S^n. If K denotes the field of fractions of R and S, then Q maps injectively into $K \underset{S}{\otimes} Q$. Fixing a finite K-basis for $K \underset{S}{\otimes} Q$ and expressing the S-generators of Q in terms of this basis, we define an embedding $Q \to S^m$ for some $m \in \mathbb{N}$. We obtained two exact sequences :

$$0 \to Q \to S^n \to S^m/Q \to 0$$

$$0 \to P \to S^n \to Q \to 0.$$

The fact that S has global dimension at most two entails that $P = P^{**}$ is projective.

IV.2.12. COROLLARY. *In the situation of the lemma : if P is κ-finitely generated, κ-closed, κ-quasiprojective then it is a projective $Q_\kappa(R)$-module.*

PROOF. Directly from Lemma III.1.16. and the above. □

IV.2.13. PROPOSITION. *Let κ_I be such that $Q_I(R)$ is Noetherian of global dimension at most two. Then there is a bijective correspondence between the class of locally projective sheaves of finite type over X(I) on one hand, and finitely generated projective $Q_I(R)$-modules on the other hand.*

PROOF. A locally projective sheaf of finite type over X(I) corresponds to a κ_I-quasiprojective κ_I-closed κ_I-finitely generated $Q_I(R)$-module, because of Proposition IV.2.5., hence to a finitely generated projective $Q_I(R)$-module (see Lemma III.1.4.). Conversely, a finitely generated projective $Q_I(R)$-module is κ_I-closed by Lemma IV.2.10. and it is certainly κ_I-finitely generated and κ_I-quasiprojective. □

IV.2.14. PROPOSITION. *Let* κ_I *be such that* $Q_I(R)$ *is Noetherian of global dimension at most two. Then there is a bijective correspondence between locally separable sheaves of algebras over* $X(I)$, *and* $Q_I(R)$*-Azumaya algebras.*

PROOF. By Corollary IV.2.6. a sheaf as mentioned in the statement corresponds to a κ_I-Azumaya algebra, A say, i.e. to a finitely generated projective central $Q_I(R)$-algebra A with the property

$$Q_I(A^e) = \text{End}_R(A) = \text{End}_{Q_I(R)}(A).$$

Now A is a finitely generated projective $Q_I(R)$-module, so the same holds for A^e and therefore A^e is κ-closed by Lemma IV.2.10., i.e. $Q_I(A^e) = A^e$ and it follows that A is indeed a $Q_I(R)$-Azumaya algebra. The converse implication is similar to that in Proposition IV.2.13. □

IV.2.15. THEOREM. *Let* $X(I)$ *be an open subscheme of* $X = \text{Spec}(R)$ *such that* $Q_I(R)$ *is a Noetherian domain of global dimension at most two, then* $\text{Br}(X(I)) = \text{Br}(Q_I(R))$.

PROOF. Using the foregoing propositions it suffices to check that κ_I-similar $Q_I(R)$-Azumaya algebras A and B are also equivalent in $\text{Br}(Q_I(R))$. If P is a κ_I-progenerator such that :

$$Q_I(A \underset{R}{\otimes} B^0) = \text{End}_R(P) = \text{End}_{Q_I(R)}(P),$$

then the fact that both A and B are finitely generated projective $Q_I(R)$-modules entails :

$$Q_I(A \underset{R}{\otimes} B^0) = Q_I(A \underset{Q_I(R)}{\otimes} B^0) = A \underset{Q_I(R)}{\otimes} B^0. \qquad \square$$

IV.2.16. Example. Let U be an open subset of the affine plane $\mathbb{A}_k^2$ when k is an algebraically closed field. Let $\Gamma(U)$ be the ring of regular functions on U, then $\text{Br}(U) = \text{Br}(\Gamma(U))$. Indeed, let U be defined by the ideal I of $k[X,Y]$, then $\text{rad}(I) = M_1 \cap \ldots \cap M_n \cap (f)$ for certain maximal ideals $M_1, \ldots, M_n$ in $k[X,Y]$ and some $f \in k[X,Y]$. Now, for each $i \in \{1,\ldots,n\}$

we have $Q_{M_i}(k[X,Y]) = k[X,Y]$ (up to changing variables $X \mapsto X + \alpha_i$, $Y \mapsto Y + \beta_i$ with $\alpha_i, \beta_i \in k$ we may assume that $M_i = (X,Y)$ and then the statement is clear). Consequently $Q_I(k[X,Y]) = k[X,Y]_f$ and so we may apply the foregoing to derive that : $Br(U) = Br(Q_I(k[X,Y])) = Br(\Gamma(U))$.

IV.3. Extending Coherent and Quasicoherent Sheaves.

The problem is the following : let Y be a subspace of the topological space X, let $\mathcal{O}_X$ be a sheaf of rings such that $(X,\mathcal{O}_X)$ is a ringed space and consider $(Y,\mathcal{O}_X|Y)$; classify the extensions of quasicoherent sheaves over Y to quasicoherent sheaves over X. If X is Noetherian and Y is open in X then a result of A. Grothendieck's states that a coherent sheaf M over Y can be extended to a coherent sheaf over X, but note that this extension need not be the direct image i_*M where $i : Y \to X$ is the canonical inclusion. Another interesting case is obtained by taking for X any scheme and putting $Y = X^{(1)}$, the subspace of points of codimension one. All these cases are special cases of the situation considered in this section, where we focus mainly on the affine case X = Spec(R) where R is a commutative Noetherian ring.

A subset Y of X = Spec(R) is said to be a T.T.-set (torsion theoretic) if there is a kernel functor κ such that $P \in X(\kappa) = X - L(\kappa)$ if and only if $P \subset Q$ for some $Q \in Y$. We say that Y is geometrically stable or generically closed if for all $x \in X$ such that the Zariski-closure of $\{x\}$ contains some $y \in Y$, we have that $x \in Y$. For a general scheme, we say that Y is (locally) a T.T.set if for all open affine subsets $Spec(R_i)$ of X, (resp. for some covering by open affine $Spec(R_i)$ in X), the sets $Y \cap Spec(R_i)$ are T.T. sets in $Spec(R_i)$. Note that here we adopt the notation $X(\kappa)$ instead of the $K(\kappa)$ introduced earlier; in this notation $X(\kappa_I) = X(I)$ for any ideal I of R.

IV.3.1. LEMMA. *Let* X = Spec(R) *for a Noetherian ring* R, *then the following statements are equivalent :*

1. Y *is geometrically stable.*
2. $Y = X(\kappa)$ *for some idempotent kernel functor* κ *on* R-*mod.*

PROOF. Straightforward. □

For a Noetherian scheme X it follows that a subspace Y of X is geometrically stable if and only if for all open affine subsets $\mathrm{Spec}(R) \subset X$ we have that $Y \cap \mathrm{Spec}(R)$ is $X(\kappa)$ for some κ on R-mod. This works for locally Noetherian schemes X by restricting attention to a covering by Noetherian affine spaces. In the sequel all schemes are understood to be Noetherian schemes, and unless otherwise stated R is a Noetherian domain.

If V is a geometrically stable subspace of X which is contained in $Y = X(\kappa)$, then we write : $\mathcal{O}_X|Y = \mathcal{O}_Y$,

resp. $$\mathcal{Q}_Y(V) = \varinjlim_U \{\Gamma(U,\mathcal{O}_X) ; U \text{ open in } X,\ V \subset U\}.$$

It is clear that $\mathcal{Q}_Y$ is a presheaf over Y and :

IV.3.2. LEMMA. *With notations as before we have :*

$$\Gamma(Y,\mathcal{Q}_Y) = \mathcal{Q}_Y(Y) = Q_\kappa(R).$$

PROOF. Since R is a domain we may write : $\mathcal{Q}_Y(Y) = \cup\{Q_I(R),\ Y \subset X(I)\}$. Now $Y \subset X(I)$ if and only if $P \in X(\kappa)$ implies $I \not\subset P$. So if I is not contained in some $P \in X(\kappa)$ and if $s \in Q_I(R)$ then $I^n s \subset R$ with $I \in L(\kappa)$ entails $s \in Q_\kappa(R)$. Conversely, for any $s \in Q_\kappa(R)$ we may find $J \in L(\kappa)$ such that $Js \subset R$. But $J \not\subset P$ for all $P \in X(\kappa)$ implies $Y \subset X(J)$ and therefore $s \in \mathcal{Q}_Y(Y)$. □

IV.3.3. COROLLARY. *If* $Y_I = X(I) \cap Y$ *is open in* Y *then* $\Gamma(Y_I,\mathcal{Q}_I) = Q_{\kappa \vee \kappa_I}(R)$.

PROOF. Y_I is geometrically stable and $Y_I = X(\kappa \vee \kappa_I)$. Replacing Y by Y_I in the lemma then yields the result. □

Note that for a Noetherian domain R we always have

$$Q_\kappa(R) = \cap\{Q_{R-P}(R); P \in X(\kappa)\}.$$

For the induced topology on Y (called the Zariski topology of Y) every open set is of the form $X(\kappa')$ for some κ', because every open set is of the form $X(I) \cap Y$ i.e. associated to $\kappa \vee \kappa_I$ for some I. Therefore we write Y(I) for Y_I; note however that Y(I) = Y(J) can happen even with rad(I) ≠ rad(J). □

IV.3.4. <u>PROPOSITION</u>. *Let* Y(I) *be open in* Y, *then :*

$$\Gamma(Y(I), \mathcal{O}_Y) = Q_\tau(R) = \cap\{Q_{R-P}(R); P \in X(\tau)\},$$

where $\tau = \kappa \vee \kappa_I$.

<u>PROOF</u>. Associating $Q_\tau(R)$ to $Y(I) = X(\tau)$ defines a sheaf (!), hence $\mathcal{Q}_Y = \mathcal{O}_Y$ and the result follows from foregoing observations. □

IV.3.5. <u>LEMMA</u>. *If* M *and* N *are* κ*-closed modules such that* $\tilde{M}|X(\kappa) = \tilde{N}|X(\kappa)$ *then* M = N.

<u>PROOF</u>. (Should be compared to Lemma IV.2.1.). By assumption $Q_{R-P}(M) = Q_{R-P}(N)$ for all $P \in X(\kappa)$. Write $E = \pi\{Q_{R-P}(M); P \in X(\kappa)\}$. The canonical map $\varphi_M : M \to E$ has kernel $Ker(\varphi_M) \subset \kappa(M) = 0$, hence φ_M is injective. Similarly, $\varphi_N : N \to E$ is injective, so we may view N and M as submodules of E. Pick $m \in M - N$ in E and consider the exact sequence :

$$0 \to N \to N + Rm \to T \to 0.$$

If $Q_{R-P}(T) = 0$ for all $P \in X(\kappa)$ then T is κ-torsion and thus $m \in Rm \subset Q_\kappa(Rm) \subset Q_\kappa(N+Rm) = Q_\kappa(N) = N$, contradiction. Therefore, $Q_{R-P}(T) \neq 0$ for some $P \in X(\kappa)$ and for this P :

$$Q_{R-P}(N) \subsetneq Q_{R-P}(N) + Q_{R-P}(R)\overline{m},$$

with $\overline{m} = j_{R-P}(m)$, $j_{R-P} : M \to M/\kappa_{R-P}(M)$. From $Q_{R-P}(M) = Q_{R-P}(N)$ we then derive a contradiction, so M = N follows. □

Consider a quasicoherent sheaf M over Y. This implies the existence of affine open subsets $U_1,\dots,U_n$ of $X = \mathrm{Spec}(R)$ such that $Y \subset \bigcup_{i=1}^{n} U_i$ and for each $V_i = Y \cap U_i$ there is an exact sequence :

$$(*) \qquad \mathcal{O}_X^{(I)}|V_i \xrightarrow{\varphi} \mathcal{O}_X^{(J)}|V_i \to M|V_i \to 0$$

for certain index sets I and J depending on i.

Put $V_i = X(\tau_i)$, then we deduce from the above another exact sequence :

$$(**) \qquad Q_{\tau_i}(R)^{(I)} \xrightarrow{f} Q_{\tau_i}(R)^{(J)} \to N_i \to 0$$

where $N_i = \mathrm{Coker}(f)$ and f is induced by φ by taking sections over V_i. Write $M_i = Q_{\tau_i}(N_i)$. We intend to show that $M|V_i = \tilde{M}_i|V_i$ and we need the following :

IV.3.6. LEMMA. *Let κ_S be the idempotent kernel functor associated to the multiplicatively closed set S in the Noetherian domain R (i.e. $L(\kappa_S) = \{I$ ideal of $R;\ I \cap S \neq \varphi\})$, and denote $\kappa \vee \kappa_S$ by τ. For any $M \in R$-mod we have : $Q_\kappa(S^{-1}M) = S^{-1}Q_\kappa(M) = Q_\tau(M)$.*

PROOF. Easy. Since κ_S is a T-functor and $\tau > \kappa_S$ the lemma is a special case of the compatibility theorems given in the non-commutative case by F. Van Oystaeyen in [89]. See also Lemma II.13.

Now we are ready to prove the following :

IV.2.7. PROPOSITION. *Let M and V_i be as before, then $M|V_i = \tilde{M}_i|V_i$ for all i.*

PROOF. The canonical map $N_i \to M_i$ induces an isomorphism $Q_{R-P}(N_i) = Q_{R-P}(M_i)$ for all $P \in V_i$, hence $\tilde{M}_i|V_i = Q_{\tau_i}(N_i)^{\sim}|V_i = \tilde{N}_i|V_i$. Exactness of $\sim$ yields that the exact sequence (**) induces another sequence :

$$(***) \qquad (Q_{\tau_i}(R)^{\sim})^{(I)}|V_i \to (Q_{\tau_i}(R)^{\sim})^{(J)}|V_i \to \tilde{N}_i|V_i \to 0.$$

Now it suffices to note that $(Q_{\tau_i}(R)^\sim)|V_i = \tilde{R}|V_i = \mathcal{O}_X|V_i$ and to compare (*) and (**).

IV.3.8. PROPOSITION. *M extends to* $U = \bigcup_{i=1}^n U_i$.

PROOF. It is not restrictive to assume that $U_i = X(f_i)$ for some $f_i \in R$, i.e. $U_i \cong \text{Spec}(R_i)$ where $R_i = Q_{f_i}(R)$. We have to check that the $\tilde{M}_i$ introduced above (but on U_i !) glue together well, i.e. $\tilde{M}_i|U_i \cap U_j = \tilde{M}_j|U_i \cap U_j$. Write $U_{ij} = U_i \cap U_j$, then $U_{ij} = \text{Spec}(R_{f_if_j})$. Since $\Gamma(U_i,\tilde{M}_i) = M_i$ it is clear that $\Gamma(U_{ij},\tilde{M}_i) = Q_{f_j}(M_i)$. Note that $V_{ij} = U_{ij} \cap Y$ equals $X(\kappa_{f_i} \vee \kappa_{f_j} \vee \kappa) = X(\tau_{ij})$ where τ_{ij} denotes $\kappa_{f_if_j} \vee \kappa$. We now have $(M_{i,f_j})^\sim|V_{ij} = (M_{j,f_i})^\sim|V_{ij}$, so it will suffice to establish that M_{i,f_j} and M_{j,f_i} are τ_{ij}-closed and then apply the Lemma IV.3.5. But this follows immediately from Lemma IV.3.6. □

For any quasicoherent sheaf M on a geometrically stable Y in X = Spec(R) we may find U open in Spec(R) such that $i : Y \to U$ allows the construction of a quasicoherent sheaf N on U with $N|Y = M$. If $j : U \to \text{Spec}(R)$ is the canonical inclusion then the fact that U is Noetherian entails that for any quasicoherent sheaf $\mathcal{Q}$ on U, $j_*\mathcal{Q}$ is quasicoherent on X. Applying this to N we obtain $M \in R$-mod such that $j_*N = \tilde{M}$, and therefore $\tilde{M}|Y = M$. If $\varphi : Y \to X$ is the canonical inclusion, we aim to establish that φ_*M is quasicoherent on X. This will follow from :

IV.3.9. PROPOSITION. *For any quasicoherent sheaf* M *on* Y, i_*M *is a quasicoherent sheaf on* U.

PROOF. Choose $T \in R$-mod such that $M = \tilde{T}|Y$ and choose a resolution $R^{(I)} \xrightarrow{\varphi} R^{(J)} \to T \to 0$ for T. From this we obtain a resolution $\mathcal{O}_\kappa^{(I)} \to \mathcal{O}_\kappa^{(J)} \to \tilde{T} \to 0$ hence an exact sequence :

$$\mathcal{O}_Y^{(I)} \xrightarrow{\psi} \mathcal{O}_Y^{(J)} \to M \to 0.$$

Write $N = \text{Coker}(Q_\kappa(\varphi) = \text{Coker}(\Gamma(Y,\psi))$ and put $M = Q_\kappa(N)$. We claim that $\tilde{M}|U \cong i_*M$, thus finishing the proof. Let $X(f) \subset U$ for some $f \in R$. By definition :

$$i_*M(X(f)) = \Gamma(Y \cap X(f), M) = \Gamma(Y \cap X(f), \widetilde{M}|Y),$$

indeed, $\widetilde{M}|Y = \widetilde{T}|Y = M$. The canonical map (derived from sheafification)

$$M_f \to \varinjlim_{\substack{W \supset Y \cap X(f) \\ W \text{ open in } X}} \Gamma(X,\widetilde{M}) \to \Gamma(Y \cap X(f), \widetilde{M}|Y)$$

defines a sheafmorphism of $\mathcal{O}_X|U$-Modules :

$$\theta : \widetilde{M}|U \to i_*M, \text{ (defined on a basis } X(f) \subset U).$$

First note that :

$$\varinjlim_{W \supset Y \cap X(f)} \Gamma(W,\widetilde{M}) = \varinjlim_{X(\kappa \vee \kappa_f) \subset X(I)} Q_I(M) = Q_{\kappa_f \vee \kappa}(M) = M_f$$

where the latter equality follows from the fact that κ_f is a T-functor while M is κ-closed cf. Lemma IV.3.6.. But then θ induces isomorphisms $\theta_P = (\widetilde{M}|U)_P \to (i_*M)_P$ for each $P \in U$, because of the following facts :

a. $(\widetilde{M}|U)_P = Q_{R-P}(M)$

b. $(i_*M)_P = \varinjlim_{P \in X(f)} \Gamma(X(f), i_*M) = \varinjlim_{P \in X(f)} Q_f(M) = Q_{R-P}(M)$.

Consequently θ is globally an isomorphism. □

IV.3.10. COROLLARY.

1. *Let* $\varphi : Y = X(\kappa) \to X = \text{Spec } R$ *be the canonical inclusion and consider a quasicoherent sheaf* M *on* Y. *Then* φ_*M *is quasicoherent on* X.
2. *If* $Y \subset X$ *is a T.T.-set of the locally Noetherian integral scheme* X *and if* M *is a quasicoherent sheaf on* Y, *then* i_*M *(where* $i : Y \to X$*) is a quasicoherent sheaf on* X.

PROOF.

1. Obvious.
2. The assertion is a local statement, so we may reduce it to the affine case and apply 1. □

Let X be a Noetherian integral scheme and let Y be a T.T.-set of X. Let $i : Y \to X$ be the canonical inclusion, and consider a coherent sheaf M on Y. We established that i_*M is a quasicoherent sheaf on X. It follows from a result of [49] that i_*M is the union of its coherent subsheaves N_α i.e. $(i_*M)(W) = \bigcup_{\alpha \in A} N_\alpha(W)$ for all W open in X.

We proceed to show that M extends to a coherent sheaf on X, by showing that M is the restriction of one of the N_α to Y. Since X is Noetherian it may be covered by a finite number of affine Noetherian subschemes $\mathrm{Spec}(R_i)$. For each $\mathrm{Spec}(R_i)$ we look for an $\alpha_i \in A$ such that $i_*M|\mathrm{Spec}(R_i) = N_{\alpha_i}|\mathrm{Spec}(R_i)$. Since A may be inductively ordered, choose $\alpha = \sup\{\alpha_i\}$ and then M will be induced by N_α. In other words we have reduced the problem to the case where X is a Noetherian affine scheme $X = \mathrm{Spec}(R)$ say. For each $P \in \mathrm{Spec}(R)$ we have an affine neighbourhood V_P of P and an exact sequence :

$$(*) \qquad \mathcal{O}_X^n|V_P \cap Y \to \mathcal{O}_X^m|V_P \cap Y \to M|V_P \cap Y \to 0.$$

A finite number of such V_P may be used to cover Spec(R). Suppose that for each of these V_P there is an $\alpha_P \in A$ such that $M|V_P \cap Y = N_{\alpha_P}|V_P \cap Y$, then taking $\alpha = \sup\{\alpha_P\}$ we obtain again that $M = N_\alpha|Y$. So we have now reduced the problem to the following. Let $X = \mathrm{Spec}(R)$ where R is a Noetherian domain, let $Y = X(\kappa)$ be generically closed in X and let M be a coherent sheaf on Y with an exact sequence :

$$\mathcal{O}_X^n|Y \to \mathcal{O}_X^m|Y \to M \to 0.$$

Let $i : Y \to X$ be the canonical inclusion and let $i_*M = \bigcup_\alpha N_\alpha$ where the N_α are coherent sheaves on Spec(R). In this situation we have :

IV.3.11. PROPOSITION. *There is an* $\alpha \in A$ *such that* $M = N_\alpha|Y$.

PROOF. Let U be an open Noetherian set containing Y, say $U = X(I)$. The exact sequence (*) induces an exact sequence in $Q_\kappa(R)$-mod :

$$(**) \qquad Q_\kappa(R)^n \to Q_\kappa(R)^m \to N \to 0.$$

Since, for all $P \in Y$, we have $\tilde{N}_P = M_P$ i.e. $Q_{R-P}(N) = M_P$. It follows that $\tilde{N}|Y = M$. Moreover, $\tilde{N}|U = \bigcup_\alpha (\tilde{N}_\alpha|U)$ where N_α runs through the finitely generated R-submodules of N.

Write $N = \Sigma_{i=1}^s Q_\kappa(R)v_i$ and consider the canonical localization maps $N \xrightarrow[j_I]{} Q_I(N) \to Q_\kappa(N)$, writing $j_I(n_i) = v_i$, $i = 1,\ldots,s$. Note that $Q_I(R) \subset Q_\kappa(R)$ since $\kappa \geqslant \kappa_I$ and R is a domain. Taking sections of $\tilde{N}|U$ we find that $Q_I(N) = \bigcup_\alpha Q_I(N_\alpha)$. Since there are only finitely many v_i, there is an α_0 such that :

$$\sum_{i=1}^{s} Q_I(R)v_i \subset Q_I(N_{\alpha_0}) \subset Q_I(N).$$

Since $\mathrm{Ker}(j_I)$ is certainly κ-torsion ($\kappa \geqslant \kappa_I$) and since $Q_\kappa(R)^s|Q_I(R)^s$ is κ-torsion it follows that $Q_I(N)|\sum_{i=1}^s Q_I(R)v_i$ is κ-torsion or $Q_\kappa(N) = Q_\kappa(\sum_{i=1}^s Q_I(R)v_i)$.

Therefore $Q_\kappa(N_{\alpha_0}) = Q_\kappa(N)$ and the canonical inclusion $Q_I(N_{\alpha_0}) \subset Q_I(N)$ gives rise to an isomorphism $(N_{\alpha_0})_P \cong N_P$ for all $P \in Y$. Hence, $M = \tilde{N}|Y = Q_\kappa(\tilde{N})|Y = Q_\kappa(N_{\alpha_0})^\sim|Y$. The canonical map $N_{\alpha_0} \to Q_\kappa(N_{\alpha_0})$ yields a morphism $\tilde{N}_{\alpha_0} \to Q_\kappa(N_{\alpha_0})^\sim$ which is an isomomorphism at each P of Y, so $\tilde{N}_{\alpha_0}|Y = Q_\kappa(N_{\alpha_0})^\sim|Y$. Finally, $M = \tilde{N}_{\alpha_0}|Y$ follows. □

Let R be a Noetherian domain and let Y be a T.T.-set in Spec(R) = X. Denote by $\hat{Y}$ the geometrically stable subspace of X associated to Y. As before we write $\mathcal{O}_Y$ for $\mathcal{O}_X|Y$ and $\mathcal{O}_{\hat{Y}}$ for $\mathcal{O}_R|\hat{Y}$. Let Q(X) be the category of quasicoherent sheaves on X.

IV.3.12. PROPOSITION. *If* $j : Y \to \hat{Y}$ *is the canonical inclusion then* $j^* : Q(\hat{Y}) \to Q(Y)$ *is an isomorphism of categories.*

PROOF. First note that in foregoing results we only used $C(\kappa) \subset Y \subset X(\kappa)$. If Y is a T.T.-set then these conditions are obviously satisfied. Therefore quasicoherent sheaves on Y extend to quasicoherent sheaves on $\hat{Y}$ (first to X and then by restriction to $\hat{Y}$). The rest of the statement is just straightforward verification.

This paragraph deals with the following problem : if M and N are quasicoherent on X, when does it happen that $M|Y = N|Y$? Let us first deal with the affine case :

IV.3.13. LEMMA. *Let* Y *be a generically closed subset of* X = Spec(R), *where* R *is a Noetherian domain. Let* $M_1, M_2 \in$ R-*mod, then the following statements are equivalent :*

1. $\tilde{M}_1|X(\kappa) \cong \tilde{M}_2|X(\kappa)$ where $Y = X(\kappa)$.
2. $Q_\kappa(M_1) \cong Q_\kappa(M_2)$.

PROOF. The localization morphism $j_\kappa : M \to Q_\kappa(M)$ induces an isomorphism $\tilde{M}|X(\kappa) \to Q_\kappa(M)^\sim|X(\kappa)$ for each $M \in$ R-mod. So 1. implies 2. is a consequence of Lemma IV.3.5. Conversely, $Q_\kappa(M_1) \cong Q_\kappa(M_2)$ yields an isomorphism $Q_\kappa(M_1)^\sim \cong Q_\kappa(M_2)^\sim$ while $Q_\kappa(M_i)^\sim|X(\kappa) = \tilde{M}_i|X(\kappa)$ for i = 1,2, so this finishes the proof.

Considering the general case of a geometrically stable Y in a Noetherian scheme X we first note that, if Spec(R) is an affine scheme open in X (R is a Noetherian domain) then Y ∩ Spec(R) is geometrically stable and we will write $Y \cap \mathrm{Spec}(R) = X(\kappa)$ for some κ. If M is a quasicoherent sheaf on X then $M|\mathrm{Spec}(R)$ is quasicoherent on Spec(R) i.e. $M|\mathrm{Spec}(R) = \tilde{M}$ for some R-module M. Assigning to each open affine of the form Spec(R) the quasicoherent sheaf of modules $\kappa(M)^\sim$ with κ as above, defines a quasicoherent sheaf on X which we will denote by $\kappa_Y M$. Indeed, as X is separated, open affines intersect in open affines and the compatibility conditions for kernel functors (cf. [89]) imply that the $\kappa(M)^\sim$ may be glued together as desired. In a similar way we obtain the quasicoherent sheaf $Q_Y(M)$ on X by assigning $Q_\kappa(M)^\sim$ to open affines of the form Spec(R).

IV.3.14. PROPOSITION. *Consider quasicoherent sheaves* M_1 *and* M_2 *on* X; *the following statements are equivalent :*

1. $M_1|Y \cong M_2|Y$.
2. $Q_Y(M_1) \cong Q_Y(M_2)$.

PROOF. Easy modification of Lemma IV.3.13. of which this is the global version. □

Let us now point out some applications. Recall that the Picard group of any ringed space $(X,\mathcal{O}_X)$ is defined in terms of isomorphism classes $[M]$ of locally free sheaves M of rank one, i.e. invertible sheaves. For such a sheaf M there exists another sheaf N of the same type such that $M \otimes N \cong \mathcal{O}_X$. Actually we may take $N \cong \mathrm{Hom}_{\mathcal{O}_X}(M,\mathcal{O}_X)$. The set of isomorphism classes of invertible sheaves of $\mathcal{O}_X$-modules will be denoted by $\mathrm{Pic}(X,\mathcal{O}_X)$ or $\mathrm{Pic}(X)$ and we may define a group structure on $\mathrm{Pic}(X)$ by introducing the multiplication $[M].[N] = [M \underset{\mathcal{O}_X}{\otimes} N]$. Recall further that a coherent sheaf of $\mathcal{O}_X$-modules, M say, is locally free (of rank n) if and only if for each $x \in X$, the $\mathcal{O}_{X,x}$-module M_x is free (of rank n).
Again we assume that X is a Noetherian integral scheme, while $Y \subset X$ is a geometrically stable subspace. Consider coherent sheaves M and N on Y and let M',N' be coherent sheaves on X with $M|Y = M$ and $N'|Y = N$. Since the sheafification functor commutes with the restriction functor, it follows that :

$$(M' \underset{X}{\otimes} N')|Y = M \underset{Y}{\otimes} N.$$

Therefore, if $M \underset{Y}{\otimes} N \cong \mathcal{O}_Y$ then it follows that :

$$(M' \underset{X}{\otimes} N')|Y = M \underset{X}{\otimes} N \cong \mathcal{O}_Y = \mathcal{O}_X|Y,$$

and the foregoing results imply that $Q_Y(M'_X \underset{X}{\otimes} N') \cong Q_Y(\mathcal{O}_X)$. Coherent sheaves of $\mathcal{O}_X$-modules M' such that for some coherent sheaf N' we have that $Q_Y(M' \underset{X}{\otimes} N') \cong Q_Y(\mathcal{O}_X)$, are said to be κ_Y-invertible sheaves.

IV.3.15. LEMMA. *The set of isomorphism classes of κ_Y-closed κ_Y-invertible sheaves of $\mathcal{O}_X$-modules forms a group* $\mathrm{Pic}(X,\kappa_Y)$.

PROOF. By definition a κ_Y-closed M is one such that $Q_Y(M) \cong M$. The multiplication in $\mathrm{Pic}(X,\kappa_Y)$ will be given by $[M_1].[M_2] = [Q_Y(M_1 \underset{Y}{\otimes} M_2)]$. The statement follows from straightforward verification; the only crucial observation is the fact that for any coherent sheaves M and N of $\mathcal{O}_X$-modules we have : $Q_Y(M \underset{X}{\otimes} Q_Y(N)) \cong Q_Y(M \underset{X}{\otimes} N) \cong Q_Y(Q_Y(M) \underset{Y}{\otimes} N)$.

This claim may be checked as follows.

First reduce to the affine case. If U is an open affine of X, say $U = \mathrm{Spec}(R)$, let $Y \cap U = X(\kappa)$ be the induced subset of Y. Since M and N are coherent, $M|U = \widetilde{M}$, $N|U = \widetilde{N}$ for some $M, N \in R$-mod and then :

$Q_Y(N)|U = Q_\kappa(N)^\sim$, hence $M \underset{X}{\otimes} Q_Y(N)|U = \widetilde{M} \underset{R}{\otimes} Q_\kappa(N)^\sim = (M \underset{R}{\otimes} Q_\kappa(N))^\sim$ and :

$Q_Y(M \underset{Y}{\otimes} Q_Y(N))|U = (Q_\kappa(M \underset{R}{\otimes} Q_\kappa(N)))^\sim$.

Similarly : $Q_Y(Q_Y(M) \otimes N|U = (Q_\kappa(Q_\kappa(M) \underset{R}{\otimes} N))^\sim$ and : $Q_Y(M \underset{X}{\otimes} N)|U = (Q_\kappa(M \underset{R}{\otimes} N)|^\sim$.

The proof of the claim now reduces to the verification of :

$$(0) : \qquad Q_\kappa(Q_\kappa(M) \underset{R}{\otimes} N) \cong Q_\kappa(M \underset{R}{\otimes} N) \cong Q_\kappa(M \underset{R}{\otimes} Q_\kappa(N)).$$

Localizing the canonical morphisms :

$$\begin{array}{ccccc} & & Q_\kappa(M) \otimes N & & \\ & \nearrow & & \searrow & \\ M \otimes N & & & & Q_\kappa(M) \otimes Q_\kappa(N) \\ & \searrow & & \nearrow & \\ & & M \otimes Q_\kappa(N) & & \end{array}$$

at any $P \in X(\kappa)$, then we see that all terms localize to $Q_{R-P}(M) \underset{Q_{R-P}(R)}{\otimes} Q_{R-P}(N)$. Since the objects in (0) are κ-closed the above isomorphisms exist. Note also that it is just as easy to verify that $[M]^{-1} = [\mathcal{Hom}_{\mathcal{O}_X}(M.\mathcal{O}_X)]$. □

IV.3.16. <u>PROPOSITION</u>. $\mathrm{Pic}(X, \kappa_Y) = \mathrm{Pic}(Y)$.

<u>PROOF</u>. Straightforward. □

IV.3.17. <u>Examples</u>.

1. Let $Y \subset X$ be an open subspace, then $\mathrm{Pic}(Y) = \mathrm{Pic}(X, \kappa_Y)$.

   If $U = \mathrm{Spec}(R)$ is an open affine in X then $Y \cap U = U(I)$ for some ideal I of R. The extensions $\widetilde{I}^m$ to X glue together well to yield a coherent sheaf $\mathcal{I}$ of $\mathcal{O}_X$-ideals. We claim that $Q_Y(M) = \varinjlim_n \mathcal{Hom}_{\mathcal{O}_X}(\mathcal{I}^n, M)$. This may be verified by reduction to the affine case, where

$Hom_{O_X}(I^n,M) = (Hom_R(I^n,M))^{\sim}$ in combination with the fact that $\varinjlim$ and $\sim$ commute.

2. Consider $Y = X^{(1)}$ the points of codimension one in a normal integral scheme X. Then : $Pic(X,\kappa_Y) = Pic(X^{(1)}) = Cl(X)$. This may be checked directly either by observing that the invertible sheaves on $X^{(1)}$ yield Cartier divisors, or in a more elementary way by reducing the first equality to the affine case where verification of $Pic(R,\kappa_1) = Cl(R)$ for any Noetherian integrally closed domain is easy. (See Section V for more detail). Note that reflexive sheaves of rank one modules on X correspond to invertible sheaves of modules on $X^{(1)} = Y$. Consequently, there is a one-to-one correspondence between the κ_Y $(=\kappa_{X^{(1)}})$-closed κ_Y-invertible modules on X and the latter sheaves. Now this is exactly the intrinsic description of reflexive sheaves given by R. Treger in [84]. However his result in loc. cit. is stated in terms of quotient categories obtained by dividing out the Serre subcategory consisting of all (quasi)coherent sheaves on X which restrict to 0 over $X^{(1)}$. Using the well-known equivalences for kernel functors and localizing subcategories it is not hard to convince oneself that both descriptions are identical. The distinction between quasicoherent and coherent is not real in this set up because of the following

IV.3.18. <u>Remark</u>. M is κ_Y-invertible if and only if M is quasicoherent and there is another sheaf of O_X-modules N of the same type such that $Q_Y(M \otimes N) \cong Q_Y(O_X)$.

<u>PROOF</u>. It suffices to establish that the assumptions entail that M is coherent. If we reduce the statement to the affine case then we only have to check that $Q_\kappa(M \otimes_R N) = Q_\kappa(R)$ entails that M is finitely generated but this is well-known, cf. Chapter II.

IV.3.19. <u>Remark</u>. For a geometrically stable Y in the Noetherian scheme X and $L \in Pic(X,\kappa_Y)$ there exists an open set U of X, $Y \subset U$, such that $Q_U(L) \in Pic(X,\kappa_U)$.

IV.4. A Note on Some Results of G. Horrocks.

In [110], G. Horrocks studies the following problem. Let $X = \mathrm{Spec}(A)$ where A is a Noetherian local ring with maximal ideal m and let $Y = X - \{m\}$ be the punctured spectrum. Classify bundles (i.e., locally free coherent sheaves) on Y. Let QS(-) be the category of quasicoherent sheaves over the space (-). Consider a subspace Y as above and let $i : Y \to X$ be the canonical inclusion. For each $F \in QS(X)$ there is a canonical transformation $\mu_F : F \to i_* i^* F$. A result of G. Horrocks states that an A-module E appears as a module of global sections of a quasicoherent sheaf over Y if and only if $\mu_{\tilde{E}}$ is an isomorphism. Write $k = A/m$ for the residue field of A, then Theorem 1.3. in [ ] states:

IV.4.1. THEOREM. *An* A-*module* E *is the module of global sections of a quasicoherent sheaf over* Y *exactly then when* $\mathrm{Hom}_A(k,E) = \mathrm{Ext}^1_A(k,E) = 0$.

We aim to show how this result may be derived directly from the relative techniques we introduced. Consider κ_m, the kernel functor with $L(\kappa_m) = \{J \text{ ideal of } A;\ J \supset m^n \text{ for some } n\}$. We obviously have a monomorphism $\mathrm{Hom}_A(k,E) \hookrightarrow \varinjlim_n \mathrm{Hom}_A(A/m^n,E)$. On the other hand $\kappa_m(E) = \{e \in E;\ m^n e = 0 \text{ for some } n \in \mathbb{N}\}$, so it is easy enough to verify that :

$$\kappa_m(E) = \varinjlim_n \mathrm{Hom}_A(A/m^n,E).$$

Now for any $J \in L(\kappa_m)$ the exact sequence

$$0 \to J \to A \to A/J \to 0$$

induces an exact sequence :

$$0 \to \varinjlim_J \mathrm{Hom}_A(A/J,E) \to E \to \varinjlim_J \mathrm{Hom}_A(J,E) \to \varinjlim_J \mathrm{Ext}^1_A(A/J,E) \to 0$$

Now, $\varinjlim_J \mathrm{Hom}_A(J,E) = Q_m(E)$ (see Section I.1., or [44]). So we see that E is κ_m-torsion free exactly then when $\varinjlim_{J \in L(\kappa_m)} \mathrm{Hom}_A(A/J,E) = 0$, i.e. when $\mathrm{Hom}_A(k,E) = 0$.

On the other hand, the second exact sequence yields that $E = Q_m(E)$, i.e.

E is κ_m-closed exactly then when $\varinjlim_{J \in L(\kappa_m)} \text{Ext}^1_A(J,E) = 0$, i.e. when $\text{Ext}^1_A(k,E) = 0$. Now we have established before that E appears as the module of sections over Y of some quasicoherent sheaf F over Y if and only if $E = Q_m(E)$. Consequently, by the above argumentation it is clear that we have G. Horrocks' result as a special case of this. We leave to the interested reader the verification of the claim that other results in [110] may also be considered from the "relative" point of view, let us just mention the following :

IV.4.2. Example. Let $\Lambda_Y = (-)^{\sim}|Y$ be the functor from A-mod to QS(Y) defined by $M \mapsto \tilde{M}|Y$. Then $H^i\Lambda_Y \cong \varinjlim_t \text{Ext}^i_A(m^t,-)$.

PROOF. For any kernel functor κ on A-mod, for any Noetherian ring A we may calculate the ith-derived functor of Q_κ as follows :

$$R^iQ_\kappa = \varinjlim_{J \in L(\kappa)} \text{Ext}^i_A(J,-).$$

Now, the description of the quasicoherent sheaves on Y = X(m), combined with the above formula applied to $\kappa = \kappa_m$, yields the result. □

CHAPTER V
APPLICATIONS TO INTEGRALLY CLOSED DOMAINS, REFLEXIVE MODULES ETC.

V.I. Relative Seminormalization.

Seminormality and seminormalization have been investigated by Andreotti and Norguet [4], in connection with the problem of obtaining algebraic parametrizations of moduli spaces. The usual procedure of normalizing has the disadvantage that too many points of the scheme split up in the process so that this method cannot lead to a genuine classification space. To remedy this one has to "glue together" points which were split up in the process of normalizing. A unified approach to this problem has been discussed by Traverso [83] and Tamone [82], where normalization is replaced by seminormalization. One of the main results of Traverso's, cf. [83], is the following : if R is a Noetherian Mori ring then R is seminormal in its integral closure if and only if for any finite set of indeterminates T we have : Pic(R) = Pic(R[T]). The latter problem is a natural one, even from the point of view of the theory of Picard groups of rings. Of course, we are interested to know when Pic(R,κ) = Pic(R[T],κ) holds, for some κ. So in this section we mimick techniques of Tamone [82] a.o. to obtain a condition for Pic(R,κ) = Pic(R[T],κ) to happen, in terms of relative seminormality with respect to κ. It is also clear that these results should have applications connected to seminormalization of open subschemes or geometrically stable subspaces, much like to applications mentioned in IV.4.

for the special case of the punctured spectrum of a local ring. However, we did not include all possible applications to the algebraic geometrical situation, because this would lead us to far, and moreover it is not in the scope of this book.

Throughout R denotes a commutative Noetherian reduced (without nilpotents) ring and κ will be a fixed, but arbitrary, idempotent kernel functor on R-mod. The Noetherian hypothesis is not always essential, and one may verify on several occasions that it would have been sufficient to work with certain ideals which are assumed to be finitely generated or some kernel functors which are assumed to be idempotent.

V.1.1. DEFINITION. Let S be an overring of R which is integral over R (overring means : $R \subset S \subset Q(R) = K$). The κ-seminormalization of R in S is given by :

$$N(R,S,\kappa) = \{s \in S, \text{ for all } P \in L(\kappa),\ j_P(s) \in Q_P(R) + J(Q_P(S))\}$$

where $j_P : S \to Q_P(S)$ is the canonical morphism.

V.1.2. PROPOSITION. *$N(R,S,\kappa)$ is the largest subring S' of S such that for all $P \in X(\kappa)$ there is a unique P', a prime ideal of S', lying over S such that the residue fields $\mathbb{k}_S(P) \to \mathbb{k}_{S'}(P)$ are isomorphic.*

PROOF. A modification of the similar statement in [83], we reproduce it here partially for convenience's sake. First we check that $N = N(R,S,\kappa)$ satisfies the requirements in the statement of the proposition. If $P \in \mathrm{Spec}(R)$ and supposing that P', P'' lie over P in N, then we claim that $b \in P'$ if and only if $b \in P''$. Indeed, since S is integral over R, hence over N, we may find Q', Q'' in Spec(S) lying over P', P'' resp. If $b \in P'$ then $b \in Q'$ and by assumption $j_P(b) = \alpha + \gamma$ with $\alpha \in Q_P(R)$ and $\gamma \in J(Q_P(S))$. From $b \in Q'$ it follows that $\alpha \in J(Q_P(S)$ and so we may assume $\alpha = 0$. Then $j_P(b) \in J(Q_P(S))$ and so $b \in Q'' \cap N = P''$ as asserted. Further, if $Q' \in \mathrm{Spec}(S)$ lies over P and $b \in N$ then $j_P(b) = \alpha + \gamma$ with

$\alpha \in Q_P(R)$ and $\gamma \in J(Q_P(S))$, consequently $b(Q') = \alpha(P) + \gamma(Q') = \alpha(P) \in \mathbb{k}_R(P)$, where the notation refers to images in the residue field with respect to the prime ideal mentioned between brackets. Conversely, if a subring A of S satisfies the above properties then $A \subset N$; this part of the proof is identical to that in [83] so we omit it here. □

V.1.2. DEFINITION. If $N(R,S,\kappa) = R$ then R is said to be κ-seminormal in S. It is evident from this definition that if R is κ-seminormal in S and T is an intermediate ring, then R is κ-seminormal in T.

Moreover, if $R \subset S$ and $R \subset T \subset N(R,S,\kappa)$ then $N(R,S,\kappa) = N(T,S,\overline{\kappa})$ where $\overline{\kappa}$ is the restriction of κ to T-mod.

IV.1.3. LEMMA. *If $\sigma \leqslant \tau$ are idempotent kernel functors on R-mod, and if R is τ-seminormal in S then R is σ-seminormal in S.*

PROOF. By definition $\sigma \leqslant \tau$ if and only if $L(\sigma) \subset L(\tau)$ if and only if $X(\tau) \subset X(\sigma)$. From Proposition V.1.2. it follows that $N(R,S,\sigma) \subset N(R,S,\tau)$ and then $N(R,S,\tau) = R$ implies $N(R,S,\sigma) = R$. □

V.1.4. COROLLARY. *If R is κ-seminormal in S then R is seminormal in S.*

PROOF. Take $\tau = \kappa$, σ trivial i.e. $L(\sigma) = \{R\}$ in the lemma. □

At this point recall the remarks preceding Lemma III.3.17. Since here we consider integral extensions S of R the "lying over" property for prime ideals does hold, so we may formulate the following easy lemma without proof :

V.1.5. LEMMA. *Let R be κ-seminormal in S. If τ is an idempotent kernel functor such that $\tau \geqslant \kappa$, let $\tau' = \inf\{\kappa_{S-Q};\ Q$ lies over some $P \in X(\tau)\}$ then the following properties hold :*

1. $X(\tau') = \{Q \in Spec(S),\ Q \cap R \in X(\tau)\}$.
2. $\tau' = \overline{\tau}$ *i.e. for every* $M \in$ S-*mod* $\tau'(M) = \tau({}_RM)$.

V.1.6. PROPOSITION. *Consider a chain of commutative rings* $R \subset B \subset S$ *integral over* R, *then* R *is* κ*-seminormal in* S *if* R *is* κ*-seminormal in* B *and* B *is* $\bar{\kappa}$*-seminormal in* S.

PROOF. By the foregoing lemma $P' \in X_B(\bar{\kappa})$ exactly when $P = P' \cap R \in X_R(\kappa)$, where X_B = Spec(B), X_R = Spec(R). Consider an element $a \in N = N(R,S,\kappa)$. For all $P' \in X_B(\bar{\kappa})$ we have $a_P = \alpha + \gamma$ for some $\alpha \in Q_P(R)$, $\gamma \in J(Q_P(S))$. Since $Q_{P'}(J(Q_P(S))) \subset J(Q_{P'}(S))$ we obtain $a_{P'} \in Q_{P'}(B) + J(Q_{P'}(S))$, for all $P' \in X_B(\bar{\kappa})$. Consequently, $a \in N(B,S,\bar{\kappa}) = B$ and thus $N(R,S,\kappa) \subset B$. But $N(R,S,\kappa) \subset N(R,B,\kappa) = R$, hence the statement follows. □

V.1.7. LEMMA. *Let* R *be* κ*-seminormal in* S *and let* C *be the conductor of* R *in* S, *then* C *is semiprime in* S *i.e.* $C = \mathrm{rad}_S(C)$.

PROOF. Since the conductor is a common ideal of R and S it will be sufficient to establish that $\mathrm{rad}_S(C) \subset R$. If $s \in S$ is such that $s^n \in C$ for some $n \in \mathbb{N}$ then for $P \in X(\kappa)$ we either have $C \subset P$, in which case $s_P^n \in Q_P(C) \subset J(Q_P(S))$ i.e. $s_P \in J(Q_P(S))$, or else $C \not\subset P$, but then $s_P \in Q_P(S) = Q_P(R)$! In both cases we have that $s_P \in Q_P(R) + J(Q_P(S))$ i.e. $s \in N(S,R,\kappa) = R$. □

We now focus on the so-called glueing of prime ideals. We assume that S is a module finite extension of the Noetherian ring R. Take $p \in \mathrm{Spec}(R)$ and let $P_1,\ldots,P_n$ be the finite set of prime ideals of S lying over p. We construct an intermediate ring B, $R \subset B \subset S$, which is maximal with respect to the following properties :

1°. There is exactly one $P' \in \mathrm{Spec}(B)$ such that $P' \cap R = P$,
2°. The canonical morphism $\mathbb{k}_R(P) \to \mathbb{k}_B(P')$ on the residue fields is an isomorphism. We say that B is obtained by glueing over P. Recall that, if B is obtained from S by glueing over P, then the conductor of B in S contains P' and furthermore if Q is a prime ideal of R such that $Q \not\supset P$ then the primes of B lying over Q correspond bijectively to the primes of S lying over Q. A further result from [83] we recall here is :

V.1.8. LEMMA. *Let* C *be the conductor of* R *in* S *and suppose that* C *is a semiprime ideal. If* P *is a prime ideal of* R*, minimal over* C *such that there is a unique* $P' \in \mathrm{Spec}(S)$ *lying over* P*, then the canonical map* $\mathbb{k}_R(P) \to \mathbb{k}_S(P')$ *is not surjective.* □

V.1.9. COROLLARY. *The conductor* C *of* R *in* $N(R,S,\kappa)$ *is not a semiprime ideal of* $N(R,S,\kappa)$.

PROOF. Using the fact that a ring obtained by glueing over a prime ideal of R is seminormal in S, the proof follows by glueing over a minimal prime ideal P of R containing C. If the ring B thus obtained is equal to N then this contradicts the properties of the ring obtained after glueing. Therefore $B \subset N(R,S,\kappa)$, but then B is κ-seminormal in $N(R,S,\kappa)$, a contradiction. □

V.1.10. THEOREM. *Let* S *be a module finite extension of the Noetherian ring* R. *If* R *is* κ*-seminormal in* S *then there is a sequence of rings :*

$$S = S_0 \supset S_1 \supset \ldots \supset S_n = R,$$

such that each S_{i+1} *is obtained from* S_i *by glueing over some prime ideal* $P_i \in L(\kappa)$.

PROOF. Suppose we already determined S_i. If $S_i = R$ then there is nothing to prove. Otherwise, let C_i be the conductor of R in S_i. Since R is κ-seminormal in S, C_i is semiprime in S_i by Lemma V.1.7. Let P be a minimal prime ideal of S_i over C_i. Then we define S_{i+1} by glueing over P in S_i. Let C_{i+1} be the conductor of R in S_{i+1}, then obviously $C_{i+1} \supset C_i$. Since S_{i+1} is obtained by glueing over P, P cannot be a minimal prime over C_{i+1}, hence $C_i \neq C_{i+1}$. The Noetherian condition entails that $S_n = R$, for some $n \in \mathbb{N}$. To finish the proof we show that each P used for glueing may be taken in $L(\kappa)$. Assume $P \in \mathrm{Spec}(R)$ does not contain C_i and let $Q_1, Q_2 \in \mathrm{Spec}(S_i)$ be such that $Q_1 \cap R = Q_2 \cap R = P$. Pick $q \in Q_1$; then $C_i q \subset Q_1 \cap R = P \subset Q_2$. However $C_i \not\subset P$ implies $C_i \not\subset Q_2$ and thus $Q_1 \subset Q_2$ follows. By symmetry $Q_1 = Q_2$, so there is a unique $Q \in \mathrm{Spec}(S_i)$ lying over P. On the other hand if for all $P \supset C_i$ we have that $P \notin C(\kappa)$ then κ-seminormality of R in S_i entails that there is a unique $P' \in \mathrm{Spec}(S_i)$ lying over P. In this case we see that there is unique lying over for all prime ideals P of R. So we only have to consider the case $P \not\supset C_i$. Then there is a unique $Q \in \mathrm{Spec}(S_i)$ lying over P, and $\mathbb{k}_R(P) = \mathbb{k}_{S_i}(Q)$. Indeed, if $x \in \mathbb{k}_{S_i}(Q)$ then $Ix \subset S_i/Q$ for some $I \not\subset Q$. Hence C_iIx injects into R/P and then $x \in \mathbb{k}_R(P)$ follows from $C_iI \not\subset P$. This leads to the conclusion $R = N(R,S_i,\kappa) = S_i$, which contradicts our assumptions. From this argumentation it is clear that, except for the trivial situation where no glueing is necessary, we always could choose the prime ideal P, which is the basis for the glueing, to be in $L(\kappa)$. □

V.1.11. COROLLARY. *If* R *is* κ*-seminormal in* S *and* M *is a multiplicative system of* R, *then* $M^{-1}R$ *is* κ'*-seminormal in* $M^{-1}S$, *where* κ' *is the idempotent kernel functor on* $M^{-1}R$*-mod induced by* $\kappa \vee \kappa_M$.

In the sequel of this section we aim to show how these techniques may be applied so as to obtain an intrinsic characterisation of κ-seminormality in case R is a Mori ring of Krull dimension one i.e. in the sequel R will be a commutative Noetherian semiprime (reduced) ring such that the integral closure of R in its total ring of fractions is a finitely generated R-module. For any extension B of R we let $\mathrm{Inv}_R(B,\kappa)$ be the set of R-submodules I of B such that there exists an R-submodule J of B such that $Q_\kappa(IJ) = Q_\kappa(JI) = B$. With these notations and assumptions we have :

V.1.12. PROPOSITION. *Let* T *be a finite set of indeterminates and write* $B = Q_\kappa(R)[T]$. *Suppose* $I \in \mathrm{Inv}_R(B,\kappa)$ *is an ideal of* B *such that* $I_0 = I \cap Q_\kappa(R)$ *contains a regular element of* R, *then the following statements are equivalent :*

1. $Q_\kappa(I) = Q_\kappa(I_0(Q_\kappa(R)[T]) = Q_\kappa(I_0)[T]$.
2. $IQ_P(R)[T]$ *is a principal ideal for all* $P \in X(\kappa)$.

PROOF. 1. ⇒ 2. From $Q_\kappa(I) = Q_\kappa(I_0)[T]$ it follows for each $P \in X(\kappa)$ that $Q_P(I) = Q_P(I_0)[T]$. The assumptions on I entail that there is a J in $Inv_R(B,\kappa)$ such that $Q_\kappa(IJ) = Q_\kappa(R)[T]$ (and J is obviously an ideal of B here.) For all $P \in X(\kappa)$ we obtain $Q_P(I)Q_P(J) = Q_P(R)[T]$, $Q_P(I) = Q_P(I_0)[T]$, and thus by (3.2.) of [13], $IQ_P(R)[T]$ is principal.
2. ⇒ 1. Assume that $IQ_P(R)[T]$ is principal for all $P \in X(\kappa)$. Note that $Q_\kappa(I) = \cap\{Q_P(I) \cap B; P \in X(\kappa)\}$. From (5.6.) in [13] it follows that $Q_P(I) = Q_P(I_0)[T]$ and this finishes the proof. □

V.1.13. COROLLARY. *Let* M *be a multiplicative subset of* R *consisting of regular elements and suppose that for each* $P \in X(\kappa)$ *with* $P \cap M \neq \phi$ *that* $Pic(Q_P(R)[T]) = 0$ *and* $Pic(M^{-1}A,\kappa) = 0$*, then we have an exact sequence :*

$$0 \to Pic(R,\kappa) \to Pic(R[T],\kappa) \to Pic(M^{-1}R[T],\kappa).$$

PROOF. Note first that although we have written κ everywhere we actually consider the induced $\overline{\kappa}$ on R[T]-mod and $\overline{\overline{\kappa}}$ on $M^{-1}R[T]$-mod. However in case R has Krull dimension one, as we assumed, $C(\kappa) = X(\kappa)$ holds. So for each $P \in C(\overline{\kappa})$ or $P' \in C(\overline{\overline{\kappa}})$ we have that $P \cap R$, $P' \cap R \in C(\kappa)$. Consequently the maps in the sequence are not only well defined but in checking local properties we may first localize at $R \cap P \in C(\kappa)$ and then localize further at P. It is easy to derive the exact sequence from the local information given. □

We may now prove the following relative version of the characterization of seminormality.

V.1.14. THEOREM. *Let* R *be a Mori ring of Krull dimension one, then the following statements are equivalent :*

1. R *is* κ*-seminormal (in its total ring of fractions).*
2. *For any finite set of indeterminates* T*, the canonical homomorphism* $Pic(R,\kappa) \to Pic(R[T],\kappa)$ *is an isomorphism.*

PROOF. 1. ⇒ 2. Assume that R is κ-semilocal and $R \neq \overline{R} = S$. Let c be the conductor of R in S and let $\{P_1,\ldots,P_n\}$ be the set of prime ideals of R containing c. Let M be the set of regular elements in R which are not contained in any of the prime ideals $P_1,\ldots,P_n$. Clearly $M^{-1}R$ is semilocal and $\mathrm{Pic}(M^{-1}R,\kappa) = 0$. If some $P \in \mathrm{Spec}(R)$ contains $s \in M$ then $P \not\supset c$. Thus $Q_P(R)$ is integrally closed and a result of Bass and Murthy yields : $\mathrm{Pic}(Q_P(R)[T]) = \mathrm{Pic}(Q_P(R)) = 0$. From V.1.13. we obtain an exact sequence :

$$0 \to \mathrm{Pic}(R,\kappa) \to \mathrm{Pic}(R[T],\kappa) \to \mathrm{Pic}(M^{-1}R[T],\kappa).$$

If R is integrally closed then we may choose M to be the set of all regular elements of R and argue similarly. It remains to establish : $\mathrm{Pic}(M^{-1}R[T],\kappa) = 0$. Let $R_1 = M^{-1}R$, $S_1 = M^{-1}S$, $c_1 = M^{-1}c$. Then the ideals $M^{-1}P_1,\ldots,M^{-1}P_n$ are the maximal ideals of R_1 and therefore we obtain that $c_1 \supset \mathrm{rad}(R_1) = J(R_1)$. Since $R' = R_1/c_1$ and $S' = S_1/c_1$ are reduced Artinian rings, $\mathrm{Pic}(R'[T],\kappa) = \mathrm{Pic}(S'[T],\kappa) = 0$. Moreover, since S_1 is semilocal integrally closed of dimension one and since the canonical map $\mathrm{Pic}(S_1[T];M,\kappa) \to \mathrm{Pic}(S_1[T],\kappa)$ is surjective, we may apply foregoing results to derive that $\mathrm{Pic}(S_1[T],\kappa) = 0$. Combining the exact sequences available to us at this point, we obtain the following commutative diagram :

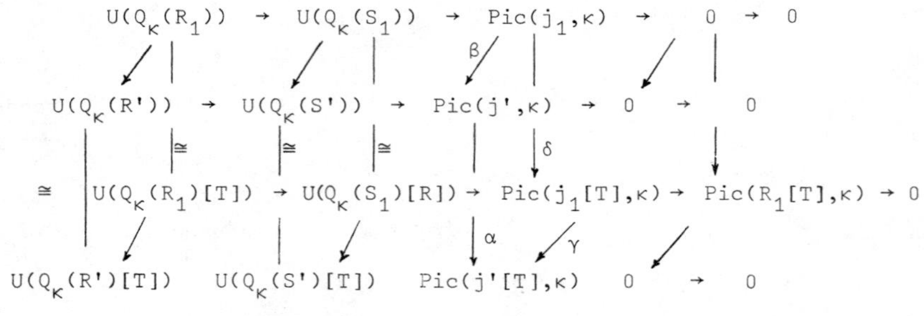

$$\begin{array}{ccccccccc}
U(Q_\kappa(R_1)) & \to & U(Q_\kappa(S_1)) & \to & \mathrm{Pic}(j_1,\kappa) & \to & 0 & \to & 0 \\
\swarrow\ | & & \swarrow\ | & & \beta\swarrow\ | & & \swarrow\ | & & \\
U(Q_\kappa(R')) & \to & U(Q_\kappa(S')) & \to & \mathrm{Pic}(j',\kappa) & \to & 0 & \to & 0 \\
\cong & & \cong & & \cong & & \downarrow\delta & & \downarrow \\
U(Q_\kappa(R_1)[T]) & \to & U(Q_\kappa(S_1)[R]) & \to & \mathrm{Pic}(j_1[T],\kappa) & \to & \mathrm{Pic}(R_1[T],\kappa) & \to & 0 \\
\cong\ \swarrow & & \swarrow & & \downarrow\alpha & & \gamma\swarrow & & \swarrow \\
U(Q_\kappa(R')[T]) & & U(Q_\kappa(S')[T]) & & \mathrm{Pic}(j'[T],\kappa) & & 0 & \to & 0
\end{array}$$

It is not hard to see that α is an isomorphism. From II.20 it follows that β and γ are isomorphisms and therefore δ is an isomorphism too. Consequently $\mathrm{Pic}(R_1[T],\kappa) = 0$.

2. ⇒ 1. Assume, on the contrary, that R is not seminormal. Put $S = N(R,\overline{R},\kappa)$ and let c be the conductor of R in S. Write $R' = R/c$, $S' = S/c$ as above and consider the square :

$$\begin{array}{ccc} Pic(R,\kappa) & \to & Pic(R[T],\kappa) \\ \downarrow & & \downarrow \\ Pic(S,\kappa) & \to & Pic(S[T],\kappa) \end{array}$$

Since S is κ-seminormal, the bottom morphism is an isomorphism. The fact that all appearing rings are reduced rings entails that,

$$U(Q_\kappa(R)) = U(Q_\kappa(R)[T]) \quad \text{and} \quad U(Q_\kappa(S)) = U(Q_\kappa(S)[T]).$$

We are left to prove that the morphism $Pic(j,\kappa) \to Pic(j[T],\kappa)$ is not an isomorphism. Now this comes down to proving that the morphism

$$Pic(j',\kappa) \to Pic(j'[T],\kappa),$$

is not an isomorphism. The proof of the latter claim is verbatim the same as in Theorem 3.6. of [83].

V.2. Noetherian Integrally Closed Domains.

Throughout R will be a Noetherian integrally closed domain with field of fractions K. We write X = Spec(R), $X^{(1)} = \{P \in X;\ htP = 1\}$. If E and F are finitely generated reflexive R-modules then $E \perp F$ stands for $(E \otimes_R F)^{**}$.
From Yuan [103], we recall that ⊥ satisfies properties similar to the usual tensorproduct and that $E^* \perp E = Hom_R(E,E)$, for E as above.
We define the rank of a reflexive R-module E to be the K-dimension of $E \otimes_R K$. So if E is a reflexive finitely generated R-module of rank one then $E^* \perp E = R$.

Following Yuan, [103], we define Γ(R), the <u>rank one reflexive class-group of R</u> to be the group of isomorphism classes of rank one reflexive

R-modules with product $[E].[F] = [E \perp F]$. It is obvious that $[R]$ is an identity element for $\Gamma(R)$ and the inverse of a class $[E]$ is given by $[E^*]$ by the remarks above.

If A is an R-algebra which is reflexive as an R-module then we say that A is an iv-R-algebra if the canonical map $A \perp A^0 \to \mathrm{End}_R(A)$ is an isomorphism. The iv-R-algebras A and B are said to be similar if there exists a pair of finitely generated reflexive R-modules E and F together with an isomorphism

$$A \perp \mathrm{End}_R(E) \cong B \perp \mathrm{End}_R(F).$$

Similarity is an equivalence relation and the classes from a group $\beta(R)$ with respect to the modified tensorproduct $\perp$. Let us now relate these notions to the relative Brauer group introduced in Chapter III.

Write $\sigma = \inf\{\kappa_{R-P};\ P \in X^{(1)}\}$.

V.2.1. LEMMA. *We have* $C(\sigma) = X^{(1)}$.

PROOF. For every prime ideal P of R either $P \in L(\sigma)$ or $\sigma(R/P) = 0$. If $P_1 \in C(\sigma)$ then $\sigma(R/P_1) = 0$, implies $\kappa_{R-P}(R/P_1) = 0$ for some $P \in X^{(1)}$. It follows that P_1 has height one. Conversely, if $P \in X^{(1)}$ then $\kappa_{R-P}(R/P) = 0$ yields $P \notin L(\sigma)$. If $P \subsetneq Q$ and $Q \notin L(\sigma)$ then argumentation as in the first part leads again to a contradiction. □

V.2.2. LEMMA. *If* M *is a torsion free* R-*module,* τ *any kernel functor on* R-*mod, then* $Q_\tau(M) = \cap\{Q_{R-P}(M);\ P \in C(\tau)\}$ *where the intersection is taken in* $K \otimes_R M$. *Consequently* M *is divisorial if and only if it is* σ-*closed.*

PROOF. The first observation follows from the fact that for M torsion free

$Q_\tau(M) = \{m \in M \otimes_R K;\ Im \subset M \subset M \otimes_R K,\ \text{for some } I \in L(\tau)\}$. The second statement follows from the first and the characterization of divisorial R-modules in [36]. □

V.2.3. PROPOSITION. *An* R-*module* E *is a finitely generated reflexive* R-*module if and only if it is* σ-*closed,* σ-*quasiprojective and* σ-*finitely generated.*

PROOF. If E is σ-closed, σ-quasiprojective and σ-finitely generated then E^* and E^{**} are finitely generated (because R is a Noetherian integrally closed domain it is certainly σ-closed and thus we apply Lemma III.1.14.) From Lemma III.1.16. it follows that E is reflexive, i.e., $E = E^{**}$ is reflexive and finitely generated. Conversely, assume E is finitely generated and reflexive. Since $R = Q_\sigma(R)$ is Noetherian, E is finitely presented and it follows that $Q_{R-P}(E)$ is a reflexive $Q_{R-P}(R)$-module for each $P \in X^{(1)}$. Since $Q_{R-P}(R)$ is a discrete valuation ring for each $P \in X^{(1)}$ (cf. [36]) it follows that $Q_{R-P}(E)$ is a projective $Q_{R-P}(R)$-module, i. e., E is σ-quasiprojective. Finally, since E is reflexive, it is torsion free, so $E = \cap\{Q_{R-P}(E);\ P \in X^{(1)}\}$, and E is σ-closed. □ σ-closed. □

V.2.4. THEOREM. *Let* R *be a Noetherian integrally closed domain. With notations as above we have :*

1. $\Gamma(R) = \mathrm{Pic}(R,\sigma)$.
2. *An* R-*algebra* A *is an iv-algebra if and only if it is a* σ-*Azumaya algebra.*
3. $\beta(R) = \mathrm{Br}(R,\sigma)$.

PROOF. 1. The results of Chapter II yield that the σ-closed σ-invertible R-modules are also σ-finitely generated and σ-quasiprojective, so Proposition V.2.3. implies that they are finitely generated and reflexive. Moreover such a module obviously has rank one. Conversely, if E is reflexive of rank one then E is finitely generated hence σ-finitely generated and σ-closed by Lemma V.2.2. For each $P \in X^{(1)}$, $Q_{R-P}(R)$ is a discrete valuation ring and $Q_{R-P}(E)$ is projective, for all $P \in X^{(1)}$, i.e. E is σ-quasiprojective. Finally, for E,F finitely generated reflexive R-modules we have : $E \perp F = Q_\sigma(E \underset{R}{\otimes} F)$, because for all torsion free finitely generated R-modules M we have that $Q_\sigma(M) = M^{**}$.

2. Immediate from foregoing observations.

3. Easy (using Theorem 3 in [103], or directly from the foregoing). □

An easy application of the interpretation of $\Gamma(R)$, $\beta(R)$ as relative "invariants" corresponding to the special kernel functor σ may be obtained as follows. Recall that for any kernel functor κ and any regular multiplicative subset S of R we have an exact sequence :

$$0 \to U(Q_\kappa(R)) \to U(S^{-1}Q_\kappa(R)) \to \mathrm{Pic}(R,S,\kappa) \to \mathrm{Pic}(R,\kappa) \to \mathrm{Pic}(S^{-1}R,\kappa)$$

(write $S^{-1}R$ for S in the sequence given in the remarks preceding II.17.). Recall also that $\mathrm{Pic}(R,S,\kappa)$ is here the group of κ-closed submodules M of $S^{-1}Q_\kappa(R)$ such that for an N of the same type, $Q_\kappa(MN) = Q_\kappa(R) = Q_\kappa(M \underset{R}{\otimes} N)$. If $Q_\kappa(R)$ is a Noetherian domain then M is a fractional ideal and $N = Q_\kappa(M^{-1})$. In this case let $D(R,\kappa)$ be the set of κ-closed fractional ideals of $Q_\kappa(R)$ with multiplication $I.J = Q_\kappa(IJ)$.

V.2.5. LEMMA. *Let* R *be a* κ*-closed Noetherian domain and let* S *be* $R - \{0\} = R^*$. *The following statements are equivalent :*

1. $D(R,\kappa)$ *is a group.*
2. $D(R,\kappa) = \mathrm{Pic}(R,R^*,\kappa)$.
3. *For all* $P \in C(\kappa)$, $Q_{R-P}(R)$ *is a discrete valuation ring of* K, *the field of fractions of* R.

PROOF. The equivalence of 1 and 2 is easy. Now note that $Q_\kappa(IJ) = R$ for κ-closed fractional ideals I and J, entails $Q_\kappa(J) = Q_\kappa(I^{-1})$. Therefore $Q_{R-P}(I)Q_{R-P}(J) = Q_{R-P}(R)$ for all $P \in C(\kappa)$ i.e. $Q_{R-P}(I)$ is invertible. Conversely, if the latter holds for each $P \in C(\kappa)$, then I is κ-invertible since $Q_\kappa(II^{-1}) = R$. This proves that each κ-closed fractional ideal of R is κ-invertible if and only if for all $P \in C(\kappa)$ each fractional ideal of $Q_{R-P}(R)$ is invertible i.e. $Q_{R-P}(R)$ has to be a discrete valuation ring. The remaining equivalences follow easily from this observation. □

Now consider σ, as before, $S = R^*$; in this case $Q_\sigma(R) = R$ and $S^{-1}Q_\sigma(R) = K$. So $\mathrm{Pic}(S^{-1}R,\sigma) = \mathrm{Pic}(K,\sigma) = 0$. In this situation we write $D(R)$ for $D(R,\sigma)$ and $\mathrm{Prin}(R)$ for K^*/R^*. The above sequence reduces to :

$$0 \to \mathrm{Prin}(R) \to D(R) \to \Gamma(R) \to 0.$$

So we recover Theorem 1 in [103] and the classical definition of the class group of a Noetherian integrally closed domain.

Another application is obtained by taking for κ the trivial kernel functor given by $L(\kappa) = \{R\}$. Then $\mathrm{Pic}(R,\kappa) = \mathrm{Pic}(R)$ and $\mathrm{Br}(R,\kappa) = \mathrm{Br}(R)$. Write $\mathrm{Bcl}(R)$ for $\mathrm{Bcl}(R,\sigma,\kappa)$. Then $\mathrm{Bcl}(R)$ is defined by a set consisting of equivalence classes of reflexive finitely generated R-modules E such that $\mathrm{End}_R(E)$ is projective. Indeed, put $A = \mathrm{End}_R(E)$, then since A is projective, $A \otimes_R A^0$ is certainly projective too, hence reflexive. It follows that :

$$A \otimes_R A^0 = Q_\sigma(A \otimes_R A^0) = Q_\sigma(\mathrm{End}_R(E) \otimes \mathrm{End}_R(E)^0).$$

Now $\mathrm{End}_R(E)^0 = \mathrm{End}_R(E^*)$ as one easily verifies. The last term in the above equality is nothing but $\mathrm{End}_R(E) \perp \mathrm{End}_R(E^*) = \mathrm{End}_R(E \perp E^*) = \mathrm{End}_R(A)$, because $E \perp E^*$ is canonically isomorphic to $\mathrm{End}_R(E) = A$. The general exact sequence reduces to :

$$0 \to \mathrm{Pic}(R) \to \Gamma(R) \to \mathrm{Bcl}(R) \to \mathrm{Br}(R) \to \beta(R).$$

In other words we recover the exact sequence of Auslander [8], and of Childs, Garfinkel and Orzech in [25]. Note that injectivity of $\mathrm{Pic}(R) \to \Gamma(R)$ does not follow from our construction; however it is straightforward to give a direct proof for this part of the sequence. Most of the results in this section may be generalized to arbitrary Krull domains if one modifies the techniques developed by M. Orzech in [67].

We now turn to a more geometrical application. Again R will denote a Noetherian integrally closed domain, I will be a fixed (non zero) prime ideal of R. We use notations : $X^{(1)}$, σ, as before. If L is any ideal of R then we write $L^{(n)}$ for $(L^n)^{**}$ and similarly we write $L \perp M$ for $(LM)^{**}$ where M is any R-module.

V.2.6. LEMMA. *Let* M *be a reflexive finitely generated* R-*module. The following statements are equivalent :*

1. $Q_I(M) = Q_I(R)$.
2. *There exists a positive integer* n *such that* $I^n \perp M = R$.

PROOF. 1. ⇒ 2. Since M is reflexive it is also torsion free and thus M embeds canonically into $Q_I(M)$. Sor for some positive n, $I^nM \subset R$. Choose n to be minimal with the forementioned properties. We may calculate within $M \otimes K = K$. We claim that for all prime ideals $P \in X^{(1)}$, $Q_{R-P}(I^nM) = Q_{R-P}(R)$. If $P \not\supset I$ then $Q_I(M) = Q_I(R)$ yields an isomorphism $Q_{R-P}(Q_I(M)) = Q_{R-P}(Q_I(I^nM)) = Q_{R-P}(I^nM) = Q_{R-P}(Q_I(R)) = Q_{R-P}(R)$. On the other hand, if $P \supset I$ then $I = P$ since $P \in X^{(1)}$. The map $Q_{R-I}(I^nM) \to Q_{R-I}(R)$ is injective and we have :

$$I^nM \subset Q_{R-I}(I^nM) \subset IQ_{R-I}(R)$$

unless $Q_{R-I}(I^nM) = Q_{R-I}(R)$, what we want to prove.
In the first case $I^nM \not\subset I$! Indeed, otherwise $Q_{R-I}(I)^nQ_{R-I}(M) \subset Q_{R-I}(I)$, and since $Q_{R-I}(I)$ is invertible in the discrete valuation ring $Q_{R-I}(R)$, $Q_{R-I}(I)^{n-1}Q_{R-I}(M) \subset Q_{R-I}(R)$ follows. For the $P \in X^{(1)}$ different from I we have $Q_{R-P}(I)^{n-1}Q_{R-P}(M) = Q_{R-P}(M) \subset Q_{R-P}(R)$. Thus,

$$\begin{aligned} I^{n-1}M &= I^{n-1}(\cap\{Q_{R-P}(M),\ P \in X^{(1)}\}) \\ &\subset \cap\{I^{n-1}Q_{R-P}(M),\ P \in X^{(1)}\} \\ &\subset \cap\{Q_{R-P}(R),\ P \in X^{(1)}\} = R. \end{aligned}$$

The choice of n yields the claim : $I^nM \not\subset I$. Pick $x \in I^nM - I \subset R - I$, $x \in IQ_{R-I}(R)$, i.e. $gx \in I$ for some $g \in R - I$ yields a contradiction, so $Q_{R-I}(I^nM) = Q_{R-I}(R)$ and finally :

$$R = \cap\{Q_{R-P}(R);\ P \in X^{(1)}\} = \cap\{Q_{R-P}(I^nM);\ P \in X^{(1)}\} = I^n \perp M.$$

2. ⇒ 1. Let $I^n \perp M = R$ i.e. $(I^nM)^{**} = R$ and pick $P \in C(\kappa_I)$, then $I^n \not\subset P$ for all n. Thus we obtain $Q_{R-P}(R) = Q_{R-P}((I^nM)^{**}) = (Q_{R-P}(I^nM))^{**}$ where the dual is now taken in $Q_{R-P}(R)$-mod. However,

$$Q_{R-P}(I^nM) = Q_{R-P}(I^n)Q_{R-P}(M) = Q_{R-P}(M),$$

and thus,

$$Q_{R-P}(R) = (Q_{R-P}(M))^{**} = Q_{R-P}(M^{**}) = Q_{R-P}(M).$$

Taking intersections over all $P \in C(\kappa_I)$ yields : $Q_I(M) = Q_I(R)$. □

V.2.7. COROLLARY. *If* M *is a finitely generated reflexive* R*-module and* I *is a prime ideal of* R *which is not minimal then* $Q_I(M) = Q_I(R)$ *implies* $M = R$.

PROOF. If $I \notin X^{(1)} = C(\sigma)$, then $\sigma(R/I) = R/I$ and thus $I \in L(\sigma)$, and consequently $I^n \in L(\sigma)$ for all $n \in \mathbb{N}$. Choose n minimal such that $I^nM \subset R$ (this is possible by the assumption $Q_I(M) = Q_I(R)$). Then $\sigma(M/I^nM) = M/I^nM$ implies $Q_\sigma(I^nM) = Q_\sigma(M)$ i.e.

$$M = Q_\sigma(M) = Q_\sigma(I^nM) = I^n \perp M = R.$$

V.2.8. COROLLARY. *For any projective* R*-module* M *the following statements are equivalent (where* I *is as before)*

1. $Q_I(M) = Q_I(R)$.
2. *There is a positive integer* n *such that* $I^n \perp M = R$.

PROOF. For a projective R-module M we have :

$$\begin{aligned} I^n \perp M &= Q_\sigma(I^nM) = Q_\sigma(I^n \underset{R}{\otimes} M) = \cap\{Q_{R-P}(I^n \underset{R}{\otimes} M);\ P \in X^{(1)}\} \\ &= \cap\{I^n \underset{R}{\otimes} Q_{R-P}(R) \underset{R}{\otimes} M;\ P \in X^{(1)}\} \\ &= (\cap\{Q_{R-P}(I)^n;\ P \in X^{(1)}\}) \underset{R}{\otimes} M = I^{(n)} \underset{R}{\otimes} M = I^{(n)}M. \end{aligned}$$ □

Let <I> the subgroup of Pic(R) generated by the isomorphism classes of invertible ideals amongst the $I^{(n)}$, $n \in \mathbb{N}$.

V.2.9. <u>THEOREM</u>. *With notations as above, we obtain the following exact sequence :*

$$0 \to \langle I\rangle \to \mathrm{Pic}(R) \xrightarrow[\varphi_I]{} \mathrm{Pic}(R,\kappa_I).$$

<u>PROOF</u>. Observe first that $[I^{(n)}] \in \mathrm{Pic}(R)$ implies $Q_{R-I}(I^{(n)}) = Q_{R-I}(R)$ because for each $P \in C(\kappa_I)$ the canonical inclusion $I^{(n)} = (I^n)^{**} \to R$ induces an isomorphism : $Q_{R-P}(I^{(n)}) \cong Q_{R-P}(R)$. Therefore $\langle I\rangle \subset \mathrm{Ker}(\varphi_I)$, where φ_I is the canonical map. Conversely, if $[M] \in \mathrm{Ker}(\varphi_I)$ then $Q_I(M) \cong Q_I(R)$ yields that $I^{(n)}M = R$ for some positive integer n. If $N \in R\text{-mod}$ is such that $M \otimes_R N = N \otimes_R M = R$, then $N = R \otimes_R N = I^{(n)}M \otimes_R N = I^{(n)}$ i.e. $I^{(n)}$ is invertible or $[I^{(n)}] \in \langle I\rangle$. □

V.2.10. <u>COROLLARY</u>. *(cf. [12]). If f is a prime element of* R *then the canonical map* $\mathrm{Pic}(R) \to \mathrm{Pic}(Q_f(R))$ *is injective.*

<u>PROOF</u>. Put I = Rf. By the foregoing $\mathrm{Ker}(\varphi_f) = \langle Rf\rangle$. But Rf^n is a free R-module for every n, hence $\langle Rf\rangle = 0$. □

V.2.11. <u>COROLLARY</u>. *Let* I *be an invertible prime ideal of* R *then the kernel of the canonical morphism* $\mathrm{Pic}(R) \to \mathrm{Pic}(Q_I(R))$ *is generated by* $[I] \in \mathrm{Pic}(R)$. □

V.2.12. <u>Remark</u>. Since R is Noetherian, every idempotent kernel functor κ has the property that its filter $L(\kappa)$ is generated by products of prime ideals in the filter. So it is rather evident that the results mentioned before referring to κ_I may be reformulated so as to apply to arbitrary κ because $\kappa = \sup\{\kappa_I;\ I \text{ a prime ideal, } I \in L(\kappa)\}$. We leave the details of this exercise to the interested reader. To conclude this section we study certain conditions under which φ_I is also surjective.

V.2.13. <u>LEMMA</u>. *Let* κ *be arbitrary and consider a* κ*-invertible* R*-module* M*, then there exists a reflexive finitely generated* R*-module* M_1 *such that* $Q_\kappa(M_1) = M$.

PROOF. Let N be a κ-closed R-module such that $Q_\kappa(M \underset{R}{\otimes} N) = Q_\kappa(R)$. As we have remarked before : $M = \mathrm{Hom}_R(N, Q_\kappa(R))$. It follows that M is torsion free and we may calculate all localizations in $M \underset{R}{\otimes} K$. In particular, M is κ-finitely presented and there is a κ-finitely generated $M' \subset M$ such that $\kappa(M/M') = M/M'$. Put $M_1 = (M')^{**}$. Then M_1 is finitely generated and the proof will be finished if we verify that $Q_\kappa(M_1) = M$. Now,

$$\begin{aligned} Q_\kappa(M_1) &= \cap\{Q_{R-P}(M_1);\ P \in C(\kappa)\} \\ &= \cap\{Q_{R-P}(M')^{**};\ P \in C(\kappa)\} \\ &= \cap\{Q_{R-P}(M)^{**};\ P \in C(\kappa)\} \end{aligned}$$

where the duals are being taken in $Q_{R-P}(R)$-mod. On the other hand $Q_\kappa(M \underset{R}{\otimes} N) = Q_\kappa(R)$ implies $Q_{R-P}(M) \underset{R}{\otimes} Q_{R-P}(N) = Q_{R-P}(M \underset{R}{\otimes} N) = Q_{R-P}(R)$, for all $P \in C(\kappa)$. Therefore $Q_{R-P}(M)$ is an invertible, hence a free module. So $Q_{R-P}(M)^{**} = Q_{R-P}(M)$ within $M \underset{R}{\otimes} K$. We obtain :

$$Q_\kappa(M_1) = \cap\{Q_{R-P}(M),\ P \in C(\kappa)\} = Q_\kappa(M) = M. \qquad \square$$

V.2.14. COROLLARY. *If for all maximal ideals* ω *of* R *with* $I \subset \omega$ *we have that* $\mathrm{Pic}(Q_{R-\omega}(R), \kappa_I) = 0$ *then the restriction morphism* $\mathrm{Pic}(R) \to \mathrm{Pic}(R, \kappa_I)$ *is surjective.*

PROOF. Take M to be a κ_I-invertible R-module and let N be a reflexive R-module such that $Q_I(N) = M$ as in the lemma. It will suffice to prove that N is an invertible R-module. If ω is a maximal ideal of R such that $I \not\subset \omega$ then $Q_{R-\omega}(Q_I(N)) = Q_{R-\omega}(M)$ is an invertible $Q_{R-\omega}(R)$-module. If on the other hand $I \subset \omega$ then by assumption : $Q_I(Q_{R-\omega}(N)) = Q_I(Q_{R-\omega}(R))$. By Lemma V.2.6. there exists a positive integer n such that :
$I^n \perp Q_{R-\omega}(N) = Q_{R-\omega}(R)$. From this we derive that :

$$\begin{aligned} Q_{R-\omega}(R) &= I^n Q_{R-\omega}(N))^{**} = (Q_{R-\omega}(I)^n Q_{R-\omega}(N))^{**} \\ &= Q_{R-\omega}(N)^{**} = Q_{R-\omega}(N). \end{aligned}$$

Consequently $Q_{R-\omega}(N)$ is invertible in this case too. Therefore $Q_{R-\omega}(N)$ is an invertible $Q_{R-\omega}(R)$-module for every maximal ideal ω of R. Globalizing this we obtain that N is an invertible R-module which finishes the proof. □

The foregoing criterion is closely linked to the following :

V.2.15. PROPOSITION. *With notations as before, the following statements are equivalent :*

1. *For all prime ideals* $P \in L(I)$, $\mathrm{Pic}(Q_{R-P}(R),\kappa_I) = 0$.
2. $Q_{R-P}(R)$ *is parafactorial for all prime ideals* P *of* R *containing* I.

PROOF. 2. ⇒ 1. Let $M \in \mathrm{Pic}(Q_{R-P}(R),\kappa_I)$. By Lemma V.2.13. we may find a finitely generated reflexive R-module N such that $Q_I(N) = M$. Let $Y \subset X = \mathrm{Spec}(R)$ be the subspace consisting of all prime ideals P of R such that $Q_{R-P}(N)$ is not free. Then Y is closed and contained in $X - X(I)$. Moreover, $I \notin Y$ because $Q_{R-I}(R)$ is a discrete valuation ring and then $Q_{R-I}(N)$ is free because it is reflexive. Similar as in [51], and using the same terminology, it follows that (X,Y) is a parafactorial couple, hence there exists an invertible sheaf $\widetilde{M}_1 = M$ over X such that $M|(X-X^{(1)}) = \widetilde{N}|(X-X^{(1)})$. But then : $M = Q_I(N) = Q_I(M_1)$, where M_1 is an invertible R-module.

1. ⇒ 2. We mimick the proof of the similar statement in [51]. Let P contain I and suppose $I \neq P$. Since $Q_{R-P}(R)P/Q_{R-P}(R)I$ contains a regular element, depth $(Q_{R-P}(R)) \geq 2$. Consider an invertible sheaf M over $U = \mathrm{Spec}(Q_{R-P}(R)) - \{Q_{R-P}(P)\}$. A result of [51] entails that it will be sufficient to establish that M is trivial.

Again, let N be a finitely generated reflexive $Q_{R-P}(R)$-module such that $\widetilde{N}|U = M$. From the fact that $U \supset \mathrm{Spec}(Q_{R-P}(R)) = \overline{\{I\}}$ it follows that $Q_I(N)$ is a κ_I-invertible $Q_I(Q_{R-P}(R)$-module. The assumption then yields : $Q_I(N) = Q_I(Q_{R-P}(R))$, and so there exists a positive integer n such that $I^n \perp N = Q_{R-P}(R)$ in $Q_{R-P}(R)$-mod. However, N is a $Q_{R-P}(R)$-module and $I \in L(\kappa_{R-P})$, so it follows from all this that $I^n \perp N = N^{**} = N$ i.e. $N = Q_{R-P}(R)$ and so M is trivial, as asserted. □

CHAPTER VI
INVARIANTS OF GRADED RINGS

VI. Background on Graded Rings and Modules.

It is obvious that the techniques introduced in Chapter II, III and IV may be applied to many classes of rings and one may hope to obtain specific new information in each of the situations considered e.g. the results on Krull domains and Noetherian integrally closed domains in Section V.2. We aim to study graded rings in some detail here. This is certainly not an arbitrary choice. Although the main motivation for this study may be found in the applications to Brauer groups of projective varieties given in Chapter VII, the ring theoretical theory has some independent interest, even in the commutative case. However the latter claim may be best supported by the explicit constructions, using graded techniques, of new examples of maximal orders and tame orders in central simple algebras which we have included in the volume "Orders", dealing with the noncommutative theory.

For full detail on graded rings we refer to C. Năstăsescu, F. Van Oystaeyen [65], or to the more exhaustive [66]. We now give a brief survey of some elementary facts and definitions, just to establish contact with this material, but we will use the books quoted as standard references for some of the more technical properties we will need in the sequel.

Throughout this chapter, let us fix notation, as follows. Let R be a commutative $\mathbb{Z}$-graded ring i.e. $R = \oplus_n R_n$, where the additive subgroups R_n of R satisfy : $R_n R_m \subset R_{n+m}$, for all $n,m \in \mathbb{Z}$. The set $h(R) = \cup_n R_n$ is called the set of homogeneous elements of R and $x \in R_n$ is said to be homogeneous of degree n. An R-module M together with a family of additive subgroups $\{M_n; n \in \mathbb{Z}\}$ satisfying $M = \oplus_{n\in\mathbb{Z}} M_n$ and $R_m M_n \subset M_{n+m}$ for all $n,m \in \mathbb{Z}$, is called a graded R-module. An R-linear map $f : M \to N$ between graded R-modules is said to be graded of degree p if $f(M_n) \subset N_{n+p}$ holds for all $n \in \mathbb{Z}$; we write $f \in \mathrm{Hom}_p(M,N)$, and $\mathrm{Hom}(M,N) = \oplus_p \mathrm{Hom}_p(M,N)$ is a graded R-module. We denote by R-gr the category consisting of graded R-modules, with graded R-linear maps of degree zero for the morphisms. For every $n \in \mathbb{Z}$ we have a functor T_n : R-gr $\to$ R-gr, $M \mapsto M(n)$ where $M(n)$ is just the R-module M endowed with the graduation $M(n)_m = M_{n+m}$. The category R-gr is a Grothendieck category but the ring R is not a generator of R-gr; actually $\oplus_{n\in\mathbb{Z}} R(n)$ is a canonical generator for R-gr. A graded R-module is gr-free if it has a basis of homogeneous elements (note that this is stronger than being free and graded !). Let - : R-gr $\to$ R-mod, $M \mapsto \underline{M}$ be the "forgetful functor" associating to the graded R-module M the underlying R-module $\underline{M}$.

If $M \in$ R-gr is finitely generated and $N \in$ R-gr arbitrary, then

$$\underline{\mathrm{HOM}(M,N)} = \mathrm{Hom}(\underline{M},\underline{N}).$$

We now list some well-known facts, cf. [65], [66].

VI.1.1. LEMMA. *Let* R *be a* $\mathbb{Z}$*-graded ring.*

1. *The graded* R*-module* P *is a projective object of the category* R-gr *if and only if* P *is a projective* R*-module. (Corollary 3.3.7., [65].)*
2. *Let* $N \subset M$ *in* R-gr *then* N *is a graded direct summand of* M *if and only if* $\underline{N}$ *is a direct summand of* $\underline{M}$. *(Corollary 3.3.8. [65].)*
3. *Let* $N \subset M$ *in* R-gr. *If* N *is an essential subobject of* M *in* R-gr *then* $\underline{N}$ *is an essential submodule of* $\underline{M}$. *(Lemma 3.3.13., [65].)*
4. *An* $M \in$ R-gr *is gr-flat (i.e.* $M \otimes_R -$ *is an exact functor on* R-gr*) if and only if* $\underline{M}$ *is a flat* R*-module. (Proposition 3.4.5., [65].)*

5. *An* $M \in$ R-gr *is a Noetherian object of* R-gr *if and only if* $\underline{M}$ *is a Noetherian* R-*module.* *(Theorem 3.3., [65].)*

6. *If* $M \in$ R-gr *is* α-*critical in* R-gr *then* $\underline{M}$ *is either* α-*critical or* $(\alpha+1)$-*critical in* R-*mod.* *(Proposition 4.15., [65].)*

7. *If* M *is Noetherian then, if* M *has Krull dimension* α *in* R-gr *we have :* $\alpha \leq K\,\dim_{\underline{R}}\underline{M} \leq \alpha + 1$. *(Theorem 4.13., [65].)*

If I is any ideal of the $\mathbb{Z}$-graded ring R then we denote by I_g the ideal generated by $I \cap h(R)$. It is easily verified that for a prime ideal P of R, P_g is again a prime ideal and ht P = ht P_g (if $P = P_g$) or else $ht(P) = 1 + ht\,P_g$. We write $Q^g_P(R)$ or $Q^g_{R-P}(R)$ for the localization of R at the multiplicative system $h(R) \cap (R-P)$; it is clear that $Q^g_{P_g}(R) = Q^g_P(R)$.

Recall that R is said to be <u>positively graded</u> if $R_m = 0$ for all $m < 0$. In that case $R_+ = \bigoplus_{n>0} R_n$ is an ideal of R and Proj(R) denotes the set of graded prime ideals of R not containing R_+ equiped with the topology induced by the Zariski topology. Positively graded rings are of course related to the geometrical situation and we shall return to these in Chapter VII. But in the stalks of the projective varieties considered there we encounter graded rings which are certainly not positively graded and which are in fact examples of strongly graded rings. A $\mathbb{Z}$-graded ring R is said to be <u>strongly graded</u> if $R_n R_m = R_{n+m}$ holds for all $n,m \in \mathbb{Z}$, i.e. if $RR_1 = R$. Some basic facts concerning strongly graded rings are listed in the following

VI.1.2. <u>LEMMA</u>. (cf. [66]). *Let* R *be strongly* $\mathbb{Z}$-*graded, then :*

1. *The functors* $R \underset{R_0}{\otimes} - : R_0\text{-mod} \to \text{R-gr}$, *and* $(-)_0 : \text{R-gr} \to R_0\text{-mod}$, *define an equivalence of categories* $R_0\text{-mod} \cong \text{R-gr}$.
2. *Every graded ideal* I *of* R *is generated by its part of degree zero, i.e.* $I = RI_0$.
3. *Every* $M \in$ R-gr *is strongly graded i.e.* $R_n M_m = M_{n+m}$ *for all* $n,m \in \mathbb{Z}$.
4. *For each* $n \in \mathbb{Z}$, $[R_n] \in \mathrm{Pic}(R_0)$. *Consequently* R *is a flat* R_0-*algebra.*

5. R *is Noetherian if and only if* R_0 *is Noetherian and if so, then* R *is finitely generated as an* R_0*-algebra.*

6. *If* R *is a domain we may consider* R *as a graded subring of its graded ring of fractions* Q^g *obtained by localization at* $S = h(R) - \{0\}$. *We have* $Q^g \cong Q(R_0)[X,X^{-1}]$ *where* $Q(R_0)$ *is the field of fractions of* R_0 *and* X *is a variable of degree one over* $Q(R_0)$. *Moreover* $R \cong \sum_{n\in\mathbb{Z}} I^n X^n$ *for some invertible ideal* I *of* R.

By a graded R-algebra we mean an R-algebra A which also has the structure of a graded ring inducing the gradation of R i.e. such that the structural morphism $R \to A$ is graded of degree zero. If R is strongly graded then every graded R-algebra will be strongly graded too.

Let us consider a commutative $\mathbb{Z}$-graded R-algebra S. Write $S^{(n)}$ for $S \underset{R}{\otimes} \dots \underset{R}{\otimes} S$ (n times S). For R-modules M and N let $t : M \underset{R}{\otimes} N \to N \underset{R}{\otimes} M$ be the switch map given by $m \otimes n \mapsto n \otimes m$. Let $\varepsilon_0 : R \to S$ be the structure map (of degree zero) and for any $M_1,\dots,M_n \in R\text{-mod}$ define

$$\varepsilon_i : M_1 \underset{R}{\otimes} \dots \underset{R}{\otimes} M_n \to M_1 \underset{R}{\otimes} \dots \underset{R}{\otimes} M_{i-1} \underset{R}{\otimes} S \underset{R}{\otimes} M_i \underset{R}{\otimes} \dots \otimes M_n$$

by $\varepsilon_i(m_1 \otimes \dots \otimes m_n) = m_1 \otimes \dots \otimes m_{i-1} \otimes 1 \otimes \dots \otimes m_n$. If $M \in R\text{-mod}$ let us write $M_1 = S \underset{R}{\otimes} M$, $M_2 = M \underset{R}{\otimes} S$, $M_{12} = M_{11} = S \underset{R}{\otimes} S \underset{R}{\otimes} M$, ...
An $S^{(2)}$-homomorphism $g : M_1 \to M_2$ gives rise to maps

$$g_1 : M_{11} \to M_{13}, \quad g_2 : M_{11} \to M_{23}, \quad g_3 : M_{13} \to M_{23}.$$

These notations will be useful in the cohomological description of the graded Brauer groups we shall introduce later.

For further use we state the graded versions of some classical theorems. We do not give the full proofs here but we refer to the classical proofs which may be easily modified so as to apply to the graded situation.

VI.1.3. THEOREM. (cf. Morita's theorem). *Let* M *be a finitely generated faithfull projective graded* R*-module and put* $A = HOM_R(M,M)$, $M^* = HOM_R(M,R)$.

The categories of graded right A-*modules and graded left* R-*modules are equivalent and the equivalence is given by the functors :* $(-) \otimes_R M$ *and* $M^* \otimes_A (-)$.

Note. Compare also to Proposition III.1.13.

VI.1.4. COROLLARY. *Let* P *and* Q *be graded* R-*progenerators (i.e. like* M *in the above theorem) and suppose that* $\alpha : \mathrm{END}(P) \to \mathrm{END}(Q)$ *is a graded* R-*algebra isomorphism of degree zero, then there exists a graded* R-*progenerator* I *of rank one and an isomorphism* f *in* R-gr, $f : P \otimes_R I \to Q$, *such that for all* $h \in \mathrm{End}(P)$ *we have* $\alpha(h) = f(h \otimes 1)f^{-1}$. *If* (J,g) *is another choice for* (I,f) *then there exists an isomorphism* ρ *in* R-gr, $\rho : I \to J$, *such that* $\lambda f = g(1 \otimes \rho)$ *for some unit* λ *of* R *of degree zero.*

PROOF. Similar to IV.1.3. in [56]. □

VI.1.5. THEOREM. (Faithfully flat descent for graded modules). *Let* S *be a graded faithfully flat ring extension of* R *and consider* $M \in$ S-gr. *Let* $g : S \otimes_R M \to M \otimes_R S$, *determine a descent-datum for* M *in the sense of* [56], *but given by a degree zero map. There exists* $N \in$ R-gr *and an isomorphism* $\eta : N \otimes_R S \to M$ *in* S-gr *such that the following diagram of graded* $S \otimes_R S$-*modules is commutative :*

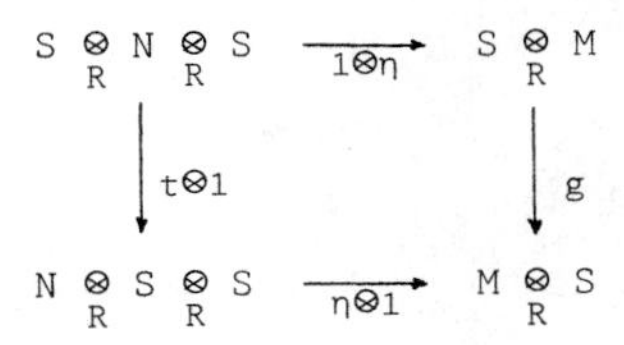

Moreover, N *is unique up to isomorphism in* R-gr.

PROOF. The classical method, cf. [56], p. 36, yields that we may take $N = \{x \in M;\ x \otimes 1 = g(1 \otimes x)\}$ and η is the map determined by $\eta(n \otimes s) = ns$ for $n \in N$, $s \in S$. By our assumptions, it is clear that N is a graded R-module and η is a graded morphism of degree zero. If (N',η') is another solution to the problem, then $\eta^{-1}\eta' : N' \underset{R}{\otimes} S \to N \underset{R}{\otimes} S$ is of the form $\alpha \otimes 1$ where $\alpha : N' \to N$ is an isomorphism, by the classical proof (loc. cit.). Obviously $\eta^{-1}\eta'$ is a graded morphism of degree zero, hence so is α. □

VI.1.6. THEOREM. (Faithfully flat descent for graded R-algebras). *In the situation of Theorem IV.1.5. we assume that* M *is a graded* S*-algebra, while* g *is supposed to be an* $S \underset{R}{\otimes} S$*-algebra isomorphism which is graded of degree zero. Then* N *has an essentially unique graded* R*-algebra structure such that* $\eta : N \underset{R}{\otimes} S \to M$ *is an* S*-algebra morphism which is graded of degree zero.*

PROOF. Formally similar to the proof of the analogous ungraded statement in [56]. □

One of the problems concerning graded rings is to detect the properties of the ring which are inherited from properties of its part of degree zero and vice-versa. The class of arithmetically graded rings is particularly well behaved from this point of view. We refer to [66], B.II for extensive treatment of these rings and we just recall some definitions and basic properties here.

A $\mathbb{Z}$-graded commutative ring R is a gr-field if each non-zero homogeneous element of R is invertible. It is easily seen that a gr-field is of the form $R_0[X,X^{-1}]$ where X is some homogeneous variable over R_0. A graded subring V of a gr-field is a gr-valuation ring if for each $x \in h(R)$ either x or x^{-1} is in V. To a gr-valuation ring there corresponds a valuation on the field of fractions K of R; the valuations of K arising in this way are called graded valuations. A gr-valuation ring is a graded discrete valuation ring if the associated valuation is discrete, i.e. the value group is $\mathbb{Z}$; this is equivalent to V being Noetherian. A graded domain R is said to be a gr-Krull domain if there exists a family of discrete gr-valuation rings $\{V_i;\ i \in I\}$ such that $R = \cap\{V_i;\ i \in I\}$ and for any $x \in h(K^g) - \{0\}$, the set $\{i \in I;\ x$ is not a unit in $V_i\}$ is a finite set (we write K^g for the gr-field of fractions of R i.e. the localization at $h(R) - \{0\}$). From [66], Proposition (B)II.1.12., it follows that a graded domain R is a gr-Krull domain if and only if it is a Krull domain. The graded class group $Cl_g(R)$ of a graded domain R is defined as $D_g(R)/P_g(R)$, where $D_g(R)$ is the subgroup of the division group D(R) consisting of graded divisorial ideals, and $P_g(R)$ is the subgroup of graded principal divisors. The graded Picard group $Pic^g(R)$ is the group of isomorphism classes (not necessarily in degree zero !) of graded projective R-modules of rank one. We let $Inv_g(R)$ be the group of invertible fractional ideals of R and then $Pic^g(R) = Inv_g(R)/P_g(R)$. For a graded Krull domain R we have $Pic^g(R) = Pic(R)$ and $Cl_g(R) = Cl(R)$. The effect of this is that we can study the Picard group and the class group of a graded Krull domain by considering graded objects only. The situation for the Brauer group is a lot more complicated and this is the topic of further sections.

Recall from [66] that a graded domain is said to be a gr-principal ideal ring if every graded ideal is principal. A graded domain R is said to be a gr-Dedekind ring if every graded ideal of R is a projective R-module. It is possible to obtain the graded versions of the usual characterizations of Dedekind rings. Let us mention some of them (see Theorem B II.2.1. in [66]).

VI.1.7. PROPOSITION. *Let* R *be a* $\mathbb{Z}$*-graded domain. The following statements are equivalent :*

1. R *is a gr-Dedekind ring.*
2. R *is a Krull domain and nonzero graded prime ideals are maximal graded ideals (we denote the latter property usually by "gr-maximal").*
3. R *is Noetherian and integrally closed in* K, *whilst nonzero graded prime ideals are gr-maximal.*
4. *Every graded ideal of* R *is invertible.*
5. *The graded fractional ideals of* R *form a multiplicative (abelian) group.*

6. R *is Noetherian and* $Q^{g}_{R-P}(R)$ *is a gr-principal ideal ring for any graded prime ideal* P *of* R. □

If R is a gr-Dedekind ring, then every fractional graded ideal of R can be generated by two homogeneous elements, one chosen arbitrarily in the ideal. Moreover, R_0 is a Dedekind ring. In case R is a strongly graded gr-Dedekind ring such that R_0 is a principal ideal ring then R is a gr-principal ideal ring and $R \cong R_0[X,X^{-1}]$ for some X, free of degree one. A subring of the form $\sum_{n\in\mathbb{Z}} I^n X^n$ of K^g, where I is an invertible ideal of R_0 is called a <u>generalized Rees ring</u> and we denote it by $\check{R}_0(I)$.

VI.1.8. <u>THEOREM.</u> *If* R *is a strongly graded gr-Dedekind ring then there exists a fractional ideal* I *of* R_0 *such that* $R = \check{R}_0(I)$. *There is a canonical group epimorphism* $\pi : Cl(R_0) \to CL(R)$; *the kernel of* π *is exactly the subgroup of* $Cl(R_0)$ *generated by the class of* I. *Moreover,* π *is an isomorphism if and only if* I *is a principal ideal and in this case* $R \cong R_0[X,X^{-1}]$. *On the other hand, for every invertible* R_0*-ideal* I, *the ring* $\check{R}_0(I)$ *is a strongly graded gr-Dedekind ring. If* $\check{R}_0(I) = \check{R}_0(J)$ *then* I *and* J *represent the same element of* $Cl(R_0)$. *Conversely, if* I *and* J *are in the same class, then* $\check{R}_0(J) = \check{R}_0(I)$.

<u>PROOF</u>. B. Theorem II.2.7. and Proposition II.2.8. in [66]. □

In the sequel R is a commutative $\mathbb{Z}$-graded domain. Let $E \in R$-gr be a finitely generated graded module and let t(E) be the submodule of E consisting of the torsion elements; obviously $t(E) \in R$-gr, and $M = E/t(E)$ is a finitely generated torsion free graded R-module. Let $\{v_1,\dots,v_n\}$ be a maximal set of homogeneous elements of M amongst some given set of homogeneous generators, such that $\{v_1,\dots,v_n\}$ is R-linearly independent. The R-module $L = Rv_1 \oplus\dots\oplus Rv_n$ is a maximal gr-free submodule of M. One may easily verify that the dimension of L is determined by M (see [ ] p. 187). The <u>graded R-length</u> (gr-length) of the finitely generated $E \in R$-gr is defined to be the dimension of L. If E is not finitely generated then the gr-length of E is defined to be the supremum of the graded lengths of finitely generated submodules of E. The graded versions of the classical properties of "length" may be derived easily (cf. [66], Lemma (B)II.2.14.) and used in proving the following theorem.

VI.1.9. <u>PROPOSITION</u>. (Graded Version of the Krull-Akizuki Theorem.) *Let* R *be a graded Noetherian domain such that every nonzero graded ideal of* R *is gr-maximal. Let* L^g *be a gr-field extension of the graded field of fractions* K^g *of* R, *such* L^g *is of finite dimension over* K^g. *Let* S *be any graded subring of* L^g *containing* R, *then* S *is Noetherian and the nonzero graded prime ideals of* S *are gr-maximal.* □

VI.1.10. <u>COROLLARY</u>. 1. *The graded integral closure of a gr-Dedekind ring in a graded extension of finite degree of its graded field of fractions is again a gr-Dedekind ring.*

2. *If* R *is a gr-Dedekind ring and* S *any multiplicatively closed subset of* $h(R) - \{0\}$. *Then* $Q_S(R)$ *is a gr-Dedekind ring.* □

Let us now include some lemmas dealing with the structure of finitely generated graded modules over gr-Dedekind rings.

VI.1.11. <u>LEMMA</u>. *Let* R *be a gr-Dedekind ring with gr-field of fractions* K^g. *If* $M \in R\text{-gr}$ *is torsion free and of rank one, then* M *is isomorphic in* R-gr *to a shifted graded fractional* R-*ideal of* K^g.

<u>PROOF</u>. Select an $m \in h(M)$ such that $K^g m = K^g M$ and consider $I = \{\alpha \in K^g;\ \alpha m \in M\}$. The latter is a graded fractional R-ideal. The map $\alpha \mapsto \alpha m$ defines a graded isomorphism $I \to M$ of degree $\deg m = d$. It is then obvious that $M \cong I(d)$ in R-gr. □

VI.1.12. <u>LEMMA</u>. *Let* R *be a gr-Dedekind ring and let* $M \in R\text{-gr}$ *be a finitely generated and torsion free graded* R-*module. If* $\alpha \in K^g$ *is such that* $\alpha M \subset M$ *then* $\alpha \in R$.

PROOF. Pick $m \neq 0$ in $h(M)$. Consider the graded fractional R-ideal $I = \{\beta \in K^g;\ \beta m \in M\}$. If $\alpha M \subset M$, then $\alpha \in I$ hence $R[\alpha] \subset I$. Since I is finitely generated in R-mod, $R[\alpha]$ is an integral extension of R in K^g. However, R is integrally closed in K^g since it is a gr-Dedekind ring, hence $\alpha \in R$. □

VI.1.13. LEMMA. *Let* R *be a gr-Dedekind ring and let* $M \in R\text{-gr}$ *be torsion free and finitely generated of rank* n. *Then* M *is graded isomorphic to* $I_1(d_1) \oplus \ldots \oplus I_n(d_n)$, *where* $d_1, \ldots, d_n \in \mathbb{Z}$ *and* $I_1, \ldots, I_n$, *are graded fractional* R*-ideals of* K^g.

PROOF. By induction on n. For $n = 1$ the statement is true because of Lemma VI.1.11.
Take $m \neq 0$ in $h(M)$. Consider $K^g m \subset K^g \underset{R}{\oplus} M$ and $K^g m \cap M$. The latter is a graded R-module and $M/K^g m \cap M$ is finitely generated and R-torsion free. Write $M \cap K^g m = N$. For all $r \in R$ we have that $rN = N \cap rM$. Indeed, suppose that $r \in h(R)$ is such that $rm_1 \in h(N \cap rM)$, with $m_1 \in M$. Then $rm_1 = \lambda M$ for some $\lambda \in K^g$ and therefore $m_1 = r^{-1}\lambda m \in N$ follows. The case $r \in R$ not necessarily homogeneous follows by using the homogeneous decomposition of r and reducing the problem to the homogeneous case by selecting the highest degree terms.
Now, M/N has rank $n - 1$, so by the induction hypothesis we obtain : $M/N \cong I_1(d_1) \oplus \ldots \oplus I_{n-1}(d_{n-1})$ while we also have a canonical graded morphism of degree zero, $\varphi : M \to I_1(d_1) \oplus \ldots \oplus I_{n-1}(d_{n-1})$. Put $M_j = \varphi^{-1}(I_j(d_j))$ and $\varphi_j = \varphi|M_j$. Then the maps $\varphi_j : M_j \to I_j(d_j)$ are graded morphisms of degree zero with $\text{Ker}(\varphi_j) = N$. Since R is a gr-Dedekind ring, each I_j is a projective graded R-module and the fact that the functors T_{d_j} are auto-equivalences of R-gr, implies that $I_j(d_j)$ is projective in R-mod. We may write $M_j = N \oplus E_j$ with $\varphi_j(E_j) = I_j(d_j)$. Obviously, the foregoing entails that

$$M = E_1 \oplus \ldots \oplus E_{n-1} \oplus N \cong I_1(d_1) \oplus \ldots \oplus I_{n-1}(d_{n-1}) \oplus N.$$

Since N has rank 1, Lemma VI.1.11 entails $N = I_n(d_n)$ for some $d_n \in \mathbb{Z}$ and some graded fractional R-ideal. □

VI.1.14. COROLLARY. 1. *Any graded finitely generated torsion free module* M *over a gr-Dedekind ring* R *is a projective* R*-module.*

2. *If* M *is as in* 1. *then* M_0 *is a finitely generated torsion free (hence projective !)* R_0*-module.*

PROOF. 1. Immediate from the foregoing and Proposition VI.1.7.4.
2. By the lemma, $M \cong I_1(d_1) \oplus \ldots \oplus I_n(d_n)$ in R-gr for some $d_1, \ldots, d_n \in \mathbb{Z}$; $I_1, \ldots, I_n$ graded fractional R-ideals. Hence $M_0 = (I_1)_{d_1} \oplus \ldots \oplus (I_n)_{d_n}$ is clearly finitely generated and torsion free as an R_0-module. Since R_0 is a Dedekind ring the statement follows. □

VI.2. Graded Picard Groups.

Throughout this section R is a commutative $\mathbb{Z}$-graded ring. We have introduced in the foregoing section the "graded Picard group" $Pic^g(R)$; but this definition was not intrinsic in R-gr because we considered (ungraded) isomorphism classes i.e. we identified certain graded modules which were not isomorphic in R-gr ! This flaw is now being dealt with, because we aim to give a better description of graded Picard groups and their interrelations in this section. We adopt the following notations and conventions. U(R) is the multiplicative group of units of R. $U_0(R)$ is the group of units of R having degree zero. $Pic^g(R)$ is the group (with respect to multiplication induced by the tensor product) of the isomorphism classes (in R-mod) of graded R-progenerators of rank one. $Pic_g(R)$ is the group (with respect to multiplication induced by the tensor product) of degree zero isomorphism classes (i.e., in R-gr !) of graded R-progenerators of rank one. G(R) is the group (multiplication is again induced by the tensor product) of isomorphism classes in R-gr of R-progenerators of rank one which are isomorphic to R as an R-module, i.e., the set underlying G(R) is the set of all possible gradations on R as an R-module. With these conventions we have:

VI.2.1. PROPOSITION. *The following sequence is exact :*

$$1 \to U_0(R) \to U(R) \xrightarrow[d_0]{} G(R) \xrightarrow[\psi]{} \mathrm{Pic}_g(R) \xrightarrow[\varphi]{} \mathrm{Pic}^g(R) \to 1,$$

where ψ *and* φ *are the natural maps and where* d_0 *is defined by assigning to any* $u \in U(R)$, *the graded* R*-module with underlying set* $\underline{R}$ *and* $(d_0(u))_i = R_i u$, *i.e. the gradation is defined by giving* u *degree zero.*

PROOF. We only check exactness at G(R). It is clear that $\psi d_0 = 1$ because for all $u \in U(R)$ the map $R \to d_0(u)$ defined by $x \mapsto xu$ is a graded isomorphism of degree zero. If $\psi(M) - 1$ then there is a graded isomorphism of degree zero $f : M \to R$. Since $\underline{M} = \underline{R}$, f is multiplication by a unit $u \in U(R)$ and so $M = d_0(u)$ follows. □

VI.2.2. LEMMA. *Let* I *and* I' *respresent the same element of* $\mathrm{Pic}^g(R)$, *i.e. there is an* R*-isomorphism* $\underline{\varphi} : I \to I'$. *There exists a unique* $T \in G(R)$ *(up to isomorphism in* R-gr*) graded* R*-module, and a graded isomorphism of degree zero* $\varphi : I \to T \otimes_R I'$ *such that the forgetful functor* $\underline{(-)} : R\text{-gr} \to R\text{-mod}$ *applied to* φ *yields* $\underline{\varphi}$. *We denote* $T = G(\varphi)$. *Then* $G(\underline{\varphi}' \circ \underline{\varphi}) = G(\varphi') \otimes_R G(\varphi)$ *and moreover if* $\underline{\varphi}$ *is multiplication by* $u \in U(R)$ *then* $G(\underline{\varphi}) = d_0(u)$.

PROOF. Consider $\varphi \otimes 1 : I \otimes_R (I')^{-1} \to R$, and define $T \in G(R)$ by putting $T_n = (\varphi \otimes 1)((I \otimes_R (I')^{-1})_n)$ for all $n \in \mathbb{Z}$. Obviously

$$\underline{\varphi} \otimes 1 : I \otimes_R (I')^{-1} \to T$$

as well as $\varphi : I \to T \otimes_R I'$ are graded morphisms of degree zero. Suppose another $T' \in G(R)$ satisfies the conditions, then the composition Φ of the following graded morphisms of degree zero :

$$T' \xrightarrow[(\varphi' \otimes 1)^{-1}]{} I \otimes_R (I')^{-1} \xrightarrow[\varphi \otimes 1]{} T$$

must be multiplication by a unit $u \in U(R)$ (because of the foregoing lemma). Forgetting the gradations, Φ becomes the identity

$$R \xrightarrow{(\varphi\otimes 1)^{-1}} I \underset{R}{\otimes} (I')^{-1} \xrightarrow{\varphi\otimes 1} R .$$

Consequently u has to be equal to 1 and therefore $T \cong T'$ in R-gr. The second statement is straightforward to verify. □

If R is a reduced ring (i.e. semiprime) then G(R) may be described completely by the idempotent elements of degree zero in R. This is due to S. Caenepeel, cf. [16].

VI.2.3. THEOREM. *Let R be a $\mathbb{Z}$-graded reduced ring.*

1. *Let* $1 = \sum_{i=1}^{n} e_i$ *be a decomposition in* R_0 *of 1 into orthogonal idempotents. For any* $\underline{d} = (d_1,\ldots,d_n) \in \mathbb{Z}^n$, *define* $M = M_{(\underline{d})}$ *in G(R) by M = R and* $M_n = R_{n-d_1}e_1 +\ldots+ R_{n-d_n}e_n$. *If* $d \neq \underline{d}'$ *then* $M_{(\underline{d})}$ *is not isomorphic to* $M_{(\underline{d}')}$ *in R-gr.*
2. *Every element of G(R) is of the form described in 1. If* $1 = e_1 +\ldots+ e_n$ *is the homogeneous decomposition of 1 as an element of* $M \in G(R)$, *then the* e_i *are homogeneous (of degree zero) orthogonal idempotent in R.*

PROOF. 1. (Here the assumption that R is reduced is not needed). If $x \in M_i \cap M_j$ with $i \neq j$ then :

$$(*) \qquad x = r_{i-d_1}e_1 +\ldots+ r_{i-d_n}e_n = s_{j-d_1}e_1 +\ldots+ s_{j-d_n}e_n$$

with $r_{i-d_1},\ldots,r_{i-d_n} \in R_{i-d_1},\ldots,R_{i-d_n}$ resp. and similarly

$$s_{j-d_1},\ldots,s_{j-d_n} \in R_{j-d_1},\ldots,R_{j-d_n}.$$

Mutliplying (*) by e_m, $m = 1,\ldots,n$, yields

$$r_{i-d_m}e_m = s_{j-d_m}e_m \in R_{i-d_m} \cap R_{j-d_m} = 0.$$

The second statement of 1. is obvious.

2. Write $e_1 = \alpha_1 +\ldots+ \alpha_m$ with $\deg_R \alpha_1 >\ldots> \deg_R \alpha_m$, $d_i \in h(R)$ where $\deg_R$ is the degree with respect to the gradation of R. Similarly, we write $\deg_M$ for the degree with respect to the gradation of M. Up to re-

ordering we may assume that $d_1 = \deg_M e_1 < \ldots < d_n = \deg_M e_n$. Up to introducing a number of zeros in $1 = e_1 + \ldots + e_n$ and in $e_1 = \alpha_1 + \ldots + \alpha_m$ we may assume that $n = m$ and $\deg_R \alpha_i = d_1 - d_i + r$, where $r = \deg_R \alpha_1 \geqslant 0$ (the zeros introduced are regarded to have the appropriate degrees to make this assumption correct !). We go on to establish $r = 0$ and $\alpha_1 = e_1$ by proving inductively the following claims :

(a) $$\alpha_1 e_2 = \alpha_1 e_3 = \ldots = \alpha_1 e_m = 0$$

(b) $$\alpha_2 e_1 = \alpha_3 e_1 = \ldots = \alpha_m e_1 = 0.$$

We prove (a), the proof of (b) is completely similar. We take parts of M-degree $r + d_m$ in :

(*) $$e_1 = e_1 . 1 = (\alpha_1 + \ldots + \alpha_m)(e_1 + \ldots + e_m)$$

we obtain $(e_1)_{r+d_m} = 0 = \alpha_1 e_m$ (note that we use the gradation of M as a graded R-module !) Suppose we have established that

$$\alpha_1 e_m = \alpha_1 e_{m-1} = \ldots = \alpha_1 e_{m-j+1} = 0.$$

Taking parts of M-degree $r + \alpha_{n-j}$ in (*) yields :

(**) $$0 = \alpha_1 e_{n-j} + \sum_k \alpha_i e_{n-k},$$

where the summation is over all k such that we have

$$d_{n-j} + r = d_1 - d_i + r + d_{n-k},$$

i.e. over the set $\{k ;\ d_i - d_1 = d_{n-k} - d_{n-j}\}$. Since $d_i - d_1 > 0$ it follows that $d_{n-k} > d_{n-j}$ i.e. $k < j$ in the summation. Multiplying (**) by $\alpha_1 e_{n-j}$ we obtain : $(\alpha_1 e_{n-j})^2 = 0$, hence $\alpha_1 e_{n-j} = 0$ (for $i \leqslant j < n - 1$). This proves (a) and (b) follows in a similar way. If $r > 0$ then $(e_1)_{d_1+r} = 0$ leads to $\alpha_1 e_1 = 0$ and $e_1 = 0$ follows then by using (b). However since we have only introduced zeros "after" e_1 and α_1 we have $e_1 \neq 0$, $\alpha_1 \neq 0$. So the foregoing reduces the problem to the case $r = 0$. In this case, taking parts of M-degree d_1 in (*) yields that

(***) $$e_1 = (e_1)_{d_1} = \alpha_1 e_1 + \ldots + \alpha_n e_m.$$

Multiplying (***) by e_1 and using (b) we obtain that $e_1^2 = \alpha_1 e_1^2$. Again from (b) it follows that $\alpha_2 e_1^2 + \alpha_3 e_1^2 + \ldots + \alpha_m e_1^2 = 0$, hence $e_1^2 = e_1^3$. Therefore e_1 is homogeneous in the gradation of R since either $\alpha_1^3 = 0$ or $\alpha_m^3 = 0$ (in the original decomposition of e_1 without zeros) in the reduced ring R yields $\alpha_1 = 0$ or $\alpha_m = 0$ etc... Consequently $\alpha_1 = e_1$ and then from (***) we get : $e_1 = e_1^2$, $e_1 e_2 = e_1 e_3 = \ldots = e_1 e_n = 0$, moreover $\deg_R e_1 = 0$.
Finally, $1 - e_1 = e_2 + e_3 + \ldots + e_n$ is the homogeneous decomposition of $1 - e_1$ in M. Write $e_2 = \beta_2 + \ldots + \beta_m$ with $\deg_R \beta_i = r + d_2 - d_i$, where we again introduce zeros (but $e_2 \neq 0$ and $\beta_2 \neq 0$) whenever necessary to make the last assumption hold. Then write :

$$e_2 = (1-e_1)e_2 = (\beta_2 + \ldots + \beta_m)(e_2 + \ldots + e_m).$$

So we may proceed exactly as before and deduce from this that e_2 is an idempotent which is homogeneous of degree zero in R and orthogonal to $e_3, \ldots, e_n$. Repetition of this argument will finally lead to the assertion of the theorem. □

For a graded ring R let $E(R)$ be the set $\{\{e_1, \ldots, e_n\}; e_i$ is idempotent in R_0 and orthogonal to e_j if $j \neq i$, $1 = \sum_i e_i\}$. The set $E(R)$ may be directed by defining $\{e_1, \ldots, e_n\} \leqslant \{f_1, \ldots, f_m\}$ to hold whenever there exists a partition $\{I_1, \ldots, I_n\}$ of $\{1, \ldots, m\}$ such that $e_i = \sum_{j \in I_i} f_j$, $i = 1, \ldots, n$. Let us write $\underline{e}$ for an element of $E(R)$. Then every $\underline{e}$, $\underline{f}$ in $E(R)$ have a successor $\underline{ef}$ which is given by $\{e_i f_j; i,j\}$. To $\underline{e} \in E(R)$ we associate $G(\underline{e}) \subset G(R)$ where $G(\underline{e})$ is the set of $M_{(\underline{d})}$ for all $\underline{d} \in \mathbb{Z}^n$, as constructed in part 1 of Theorem VI.2.3. The map $\mathbb{Z}^n \to G(\underline{e})$, $\underline{d} \mapsto M_{(\underline{d})}$ is bijective. If $\underline{e} \leqslant \underline{f}$ in $E(R)$ then $G(\underline{e}) \subset G(\underline{f})$, the latter inclusion may be described by sending $(d_1, \ldots, d_n)$ to $(c_1, \ldots, c_m)$ where $c_j = d_i$ whenever $j \in I_i$, $I_1 \cup \ldots \cup I_n = \{1, \ldots, m\}$ being the partition describing $\underline{e} \leqslant \underline{f}$ as defined above. From the theorem we deduce :

VI.2.4. <u>COROLLARY</u>. *If* R *is a* $\mathbb{Z}$*-graded commutative ring then there is an embedding* $\varinjlim_{\underline{e}} G(\underline{e}) \to G(R)$ *and this inclusion becomes the identity if* R *is reduced. In this case* G(R) *is isomorphic to the group* $\bigoplus_{i \in J} \mathbb{Z}$ *for some* *index set* J. *If* E(R) *has a maximal element* $\underline{e} = \{e_1,\ldots,e_n\}$ *then we have* $G(R) = \mathbb{Z}^n$. *If there are no nontrivial idempotents in* R *then* $G(R) = \mathbb{Z}$. □

VI.2.5. LEMMA. *Let* R *be a* $\mathbb{Z}$*-graded reduced ring without nontrivial idempotents then every unit of* R *is homogeneous.*

PROOF. If R is a domain then every unit of R is in h(R) as one easily checks. If R is not a domain then we have minimal prime ideals P_α of R, (and these are graded ideals), such that $\bigcap_\alpha P_\alpha = 0$. The units of R/P_α are homogeneous. Let u be a unit of R and assume u decomposes as $u_1 +\ldots+ u_n$ with $\deg u_i = d_i$ and $d_1 <\ldots< d_n$. Put $v = u^{-1} = v_1 +\ldots+ v_n$ with $\deg v_i = -d_i$ (up to introducing zeros in the decompositions we may always assume this to hold, by giving the required degree to each zero when necessary). We claim that

(a) $$u_n v_1 = u_n v_2 =\ldots= u_n v_{n-1} = 0$$

(b) $$u_1 v_n = u_2 v_n =\ldots= u_{n-1} v_n = 0.$$

The same method of proof as used in Theorem VI.2.3. may be used to establish this claim.

Now $(uv)_0 = u_1v_1 +\ldots+ u_nv_n = 1$. Multiplying by u_nv_n yields $u_nv_n = (u_nv_n)^2$ hence either $u_nv_n = 0$ or $u_nv_n = 1$. If $u_nv_n = 0$ then it follows from (a) and (b) that $(u_1 +\ldots+ u_{n-1})(v_1 +\ldots+ v_{n-1}) = 1$. By induction we keep on reducing the length of the decomposition of some unit in R or we obtain $u_jv_j = 1$ at some step. Hence we have reduced the problem to the situation where $u = u_1 +\ldots+ u_n$ and at least one of the homogeneous components u_j is a unit of R, while $v = v_1 +\ldots+ v_n$ with $u_jv_j = 1$. In this situation $u_j \bmod P_\alpha \neq 0$ for all α and as u mod P_α is a unit of R/P_α, u mod P_α is homogeneous. Consequently u mod $P_\alpha = u_j \bmod P_\alpha$ holds for all α, i.e. if $u_i \neq u_j$ then $u_i \in P_\alpha$ for all α and thus for $i \neq j$, $u_i \in \bigcap_\alpha P_\alpha = 0$, which proves that $u = u_j$. □

VI2.6. Remark. Another proof may be obtained from Corollary VI.2.4., Theorem VI.2.3. and Proposition VI.2.1. □

VI.2.7. Examples. The exact sequence mentioned in Proposition VI.2.1. may be explicitly calculated in several examples; let us mention some here :

a) $R = k[T]$ where k is a field and T is a variable with $\deg T \neq 0$, then :

$$1 \to k^* \to k^* \to \mathbb{Z} \to \mathbb{Z} \to 1 \to 1$$

b) $R = k[T,T^{-1}]$ with $\deg T = 1$:

$$1 \to k^* \to h(R)^* \cong k^* \oplus \mathbb{Z} \to \mathbb{Z} \to 1 \to 1 \to 1$$

c) $R = k[T,T^{-1}]$ with $\deg T = 2$:

$$1 \to k^* \to h(R)^* \cong k^* \oplus 2\mathbb{Z} \to \mathbb{Z} \to \mathbb{Z}/2\mathbb{Z} \to 1 \to 1$$

d) $R = \mathbb{C} \otimes_{\mathbb{R}} \mathbb{C}$ with trivial gradation :

$$1 \to U(R) \to U(R) \to \mathbb{Z} \oplus \mathbb{Z} \to \mathbb{Z} \oplus \mathbb{Z} \to 1 \to 1.$$

Here $G(R) = G(\underline{e})$ where $\underline{e} = \{\frac{1}{2}(1 \otimes 1 - i \otimes i),\ \frac{1}{2}(1 \otimes 1 + i \otimes i)\}$.

e) $R = \mathbb{C} \otimes_{\mathbb{R}} \mathbb{C}[T,T^{-1}]$ with $\deg T = 1$:

$$1 \to U_0(R) \to U_0(R) \oplus \mathbb{Z} \oplus \mathbb{Z} \to \mathbb{Z} \oplus \mathbb{Z} \to 1 \to 1.$$

f) $R = \mathbb{C} \otimes_{\mathbb{R}} \mathbb{C}[T,T^{-1}]$ with $\deg T = 2$:

$$1 \to U_0(R) \to U_0(R) \oplus 2\mathbb{Z} \oplus 2\mathbb{Z} \to \mathbb{Z} \oplus \mathbb{Z} \to \mathbb{Z}/2\mathbb{Z} \oplus \mathbb{Z}/2\mathbb{Z} \to 1 \to 1.$$

Note that $U(\mathbb{C} \otimes_{\mathbb{R}} \mathbb{C}) \cong \mathbb{R}^* \oplus \mathbb{R}^*$ because it is given as the set $\{ae_1 + be_2;\ a,b \in \mathbb{R}^*\}$ where $\underline{e} = (e_1,e_2)$ (as in d).

g) If R is not semiprime then Theorem 2.3. does not hold; for example take $R = \mathbb{C}[x]$ where $x^2 = 0$ and $\deg x = 1$. Then $G(R) \cong \mathbb{Z} \oplus \mathbb{C}$; for $(m,\lambda) \in \mathbb{Z} \oplus \mathbb{C}$ let $M_{(m,\lambda)}$ be $\underline{R}$ with gradation determined by giving $1 + \lambda x$ degree m. The homogeneous decomposition of 1 in $M_{(m,\lambda)}$ becomes $1 = (1+\lambda x) + (-\lambda x)$.

We consider $G(-)$ as a covariant functor from the category of graded commutative rings with graded ring morphisms of degree zero to the category of abelian groups. Let S be a graded commutative ring extension of

R, then we may define the Amitsur complex $C(S/R,G(-))$ with respect to the functor $G(-)$, cf. [56].

VI.2.8. PROPOSITION. (Faithfully flat descent for $G(-)$). *Let* S *be a graded commutative faithfully flat extension of* R, *then* $H^0(S/R,G(-)) = G(R)$. *Hence, an element* $M \in G(S)$ *is of the form* $N \otimes_R S$ *for some* $N \in G(R)$ *if and only if :* $S \otimes_R M = M \otimes_R S$ *in* $G(S \otimes S)$.

PROOF. If $N \in G(R)$, then $N \otimes_R S \in H^0(S/R,G(-))$ is clear. Suppose $M \in G(S)$ is such that $S \otimes_R M = M \otimes_R S$ in $G(S \otimes_R S)$. Then there exists a graded $S \otimes_R S$-module isomorphism of degree zero, $\varphi : S \otimes_R M \to M \otimes_R S$ which reduces to the identity if we forget gradations. Obviously φ presents us with a descent datum in degree zero, defining an R-module N which makes the following diagram of graded $S \otimes_R S$-module isomorphisms of degree zero into a commutative one :

$$\begin{array}{ccc} S \otimes_R N \otimes_R S & \xrightarrow{1\otimes\eta} & S \otimes_R M \\ \downarrow{\scriptstyle t\otimes 1} & & \downarrow{\scriptstyle \varphi} \\ N \otimes_R S \otimes_R S & \xrightarrow{\eta\otimes 1} & M \otimes_R S \end{array}$$

where t is switch morphism.

From $N \in G(R)$ it follows that $M = N \otimes_R S$ is in $G(S)$.

VI.3. Graded Azumaya Algebras and Graded Brauer Groups.

First we go back to Section III.2. and look at separable extensions of graded commutative rings. It is clear that the natural gradation on $A^e = A \otimes_R A^0$ for some $\mathbb{Z}$-graded R-algebra A has the property that the exact sequence (*) in III.2., i.e.

$$0 \to J \to A^e \xrightarrow[m_A]{} A \to 0$$

is a sequence of graded morphisms of degree zero, consequently J is a graded left ideal of A^e. We say that A is a gr-separable R-algebra if A is a projective A^e-module in R-gr. Since this implies that $\underline{A}$ is projective in A^e-mod it follows that a graded R-algebra is gr-separable if and only if it is separable. Note that a separability idempotent $\varepsilon \in A^e$ (in the sense of [56] a.o. ...), which need not be homogeneous, may be replaced by its component of degree zero because $m_A(\varepsilon) = 1$ yields $m_A(\varepsilon_0) = 1$ and $J\varepsilon = 0$ yields $J\varepsilon_0 = 0$ since J is graded, i.e. ε_0 is also a separability idempotent. The reader may easily convince himself of the validity of the graded versions of Lemma III.2.1.; one should be careful however to substitute $HOM_{A^e}(M,-)$ for $Hom_{A^e}(M,-)$ where necessary. The definition of a normal extension and a Galois extension of a graded ring remains unaltered, (see Definition III.2.2.), if one restricts attention to graded automorphisms of degree zero for the Galois automorphisms. Furthermore, one easily phrases the graded equivalents of G1,...,G5 given on p.49, in doing so one has to take care to state these properties in terms of homogeneous elements, degree zero morphisms, graded modules etc...

In the graded version of G3 it is clear that the gradation on the algebra A(C,G,1) should be defined by giving the elements u_σ, $\sigma \in G$, degree zero. The isomorphism mentioned in G3 will then become a graded isomorphism of degree zero in the graded case. It is an easy exercise to prove that the graded versions $G1^g,\ldots,G5^g$ of the properties on p.49 are still equivalent. Only concerning $G5^g$ there may be some hesitation, because the graded version of G5 is dealing with gr-maximal ideals and the behaviour of these is different from that of maximal ideals. Nevertheless one may prove $G2^g \Leftrightarrow G5^g$ in exactly the same way as the equivalence G2 $\Leftrightarrow$ G5 has been established in [68]. Finally one may also produce the graded version of Proposition III.2.4.; one has to take for γ in III.2.4.(4) a graded morphism of degree zero and it is sufficient to assume here that C does not contain <u>homogeneous</u> idempotents different from 0,1.

VI.3.1. <u>LEMMA</u>. *Let* R *be a commutative* $\mathbb{Z}$*-graded ring and let* C *be a graded Galois extension of* R.

1. *If* R *is strongly graded then* C_0 *is a Galois extension of* R_0 *with the same Galois group.*

2. *If* R *is positively graded then* $C = R \otimes_{R_0} C_0$.

PROOF. 1. Follows from the equivalence of the categories R-gr and R_0-mod applied to $G1^g, G2^g \ldots,$ or $G5^g$.
2. Since C is a projective R-module the statement follows readily from a result of T. Lam, [111]. □

The introductory remarks about gr-separability indicate that the Azumaya algebras over R which are graded R-algebras are exactly the gr-Azumaya algebras (i.e. the gr-separable R-algebras with center R). If [A] and [B] are isomorphy classes (in the graded sense !) of graded Azumaya R-algebras A and B then we define the equivalence relation $\underset{gr}{\sim}$ by saying that $[A] \underset{gr}{\sim} [B]$ if and only if there exists graded finitely generated projective R-modules P and Q such that there is an isomorphism of graded R-algebras

$$(*) \qquad A \otimes_R \mathrm{END}(P) \cong B \otimes_R \mathrm{END}(Q).$$

Since both P and Q are finitely generated, we may write End(P) for END(P) and End(Q) for END(Q). It is clear that (*) is also equivalent to the existence of a graded isomorphism,

$$(**) \qquad A \otimes_R B^0 \cong \mathrm{END}(P') = \mathrm{End}(P'),$$

for a graded finitely generated projective R-module P'. The equivalence classes for the relation $\underset{gr}{\sim}$ form a group with respect to the operation induced by the tensor product $\otimes_R$. This abelian group is called the graded Brauer group of R and it is denoted by $Br^g(R)$. The forgetful functor defines a group homomorphism $Br^g(R) \to Br(R)$, sending the class of [A] to the class of $[\underline{A}]$. Obviously it is very well possible that $A \otimes_R B^0 = \mathrm{End}(P)$ for some finitely generated projective R-module P whereas the fact that A and B and thus $A \otimes_R B^0$ are graded R-algebras need not imply that P may be taken to be a graded module. In other words, the morphism

$Br^g(R) \to Br(R)$ need not be injective. In the theory we develop it will be shown that for arithmetically graded rings this map is injective and moreover it will become clear that the graded Brauer group may be used to describe the Brauer group completely, in many cases. A commutative graded extension C of R gr-splits the graded Azumaya algebra A over R if $A \otimes_R C \cong END_C(P)$ for some finitely generated projective C-module $P \in C\text{-gr}$. Clearly C splits A if C gr-splits A but the converse need not hold in general. As usual we denote the subgroup of $Br^g(R)$, resp. $Br(R)$, consisting of the classes of algebras gr-split (resp. split) by C, by $Br^g(C/R)$, (resp. $Br(C/R)$).

VI.3.2. PROPOSITION. *If* R *is strongly graded then* $Br^g(R) = Br(R_0)$.

PROOF. Since R is strongly graded every graded Azumaya algebra over R is a strongly graded ring. If A is such an algebra then A^e will be a strongly graded ring too. Now Lemma VI.1.2.1. implies that A will be a projective A^e-module if and only if A_0 is a projective $(A^e)_0$-module. For any $n \in \mathbb{Z}$, $A_n \otimes_{R_0} A^0_{-n} = A_n R_{-n} \otimes_{R_0} R_n A^0_{-n}$ follows from $R_{-n} \otimes_{R_0} R_n = R_{-n}R_n = R_0$ (see Lemma VI.1.2.4.). Hence we have :

$$(A^e)_0 = \sum_n A_n \otimes_{R_0} A^0_{-n} = A_0 \otimes_{R_0} A^0_0 = (A_0)^e.$$

The maps $Br^g(R) \to Br(R_0)$, $[A] \mapsto [A_0]$, $Br(R_0) \to Br^g(R)$, $[A] \mapsto [A \otimes_{R_0} R]$, are invers to each other, hence the statement follows. Note that taking parts of degree zero (or conversely $- \otimes_{R_0} R$) respects the equivalence relations defining the Brauer groups involved ! □

The above proposition may give a rather negative impression; indeed it shows that the graded Brauer group may reduce to a fairly trivial object in the sense that it need not be very close to $Br(R)$. However in relating $Br^g(R)$ to $Br(R)$ we have some arbitrarity in changing the gradation of R e.g. blow up the degrees by putting $R'_{en} = R_n$ for a fixed e, for all $n \in \mathbb{Z}$. This will prove to be most useful.

If R is a graded integrally closed domain then a graded Azumaya algebra A over R will be a maximal order in $K \otimes_R A = \Sigma$ where K is the field of fractions of R. The graded ring of fractions $Q^g(A) = K^g \otimes_R A$ is a gr-c.s.a. (graded central simple algebra) over the graded field of fractions K^g of R. This brings us to the first problem i.e. determine the graded Brauer group of a gr-field. We extend the terminology introduced before Proposition IV.1.7. and say that Δ is a gr-field if every $\lambda \in h(\Delta) - \{0\}$ is invertible (so we do not say that Δ is a gr-skewfield). A gr-field which is commutative is of the form $k[T,T^{-1}]$ where k is a field and T is a variable with deg $T = t \in \mathbb{Z}$. The structure of general gr-fields is well-known too :

VI.3.3. PROPOSITION. *If Δ is a gr-field then Δ is graded isomorphic to $\Delta_0[X,X^{-1},\varphi]$ where Δ_0 is a field (non-commutative), φ is an automorphism of Δ_0 and X is a variable such that multiplication is determined by* $Xd = \varphi(d)X$ *for all* $d \in \Delta_0$.

A ring S is said to be gr-simple if it has no graded ideals different from zero. We say that the graded ring S is (left) gr-Artinian if it satisfies the descending chain condition on graded (left) ideals. Every gr-simple gr-Artinian ring A is isomorphic to some $M_n(\Delta)(\underline{d})$ where Δ is a gr-field and the gradation on the $n \times n$-matrices over Δ is defined by $d \in \mathbb{Z}^n$ as follows :

$$(M_n(\Delta)(\underline{d}))_\lambda = (\Delta_{\lambda+d_i-d_j})_{ij}, \quad \text{for } \lambda \in \mathbb{Z}.$$

For the proof of Proposition VI.3.3. as well as for the formetioned results the reader may consult [65] or [66]. As we have presented the material here, a gr-c.s.a. is just a gr-simple gr-Artinian algebra which is a finite module over its center. Actually this may be clarified by first noting that the center of a gr-simple gr-Artinian ring A is a gr-field and then one may apply

VI.3.4. PROPOSITION. *If A is a graded P.I.-algebra such that Z(A) is a gr-field then A is an Azumaya algebra over Z(A).*

PROOF. cf. [66] or [92]. Here we do not go into the theory of P.I. rings, the interested reader may consult C. Procesi's book, [71]. Let us just point out that any ring which is finitely generated over its center is a P.I. algebra, so VI.3.4. may indeed be viewed as a characterization of gr-c.s.a. as being Azumaya algebras over gr-fields. □

A direct application of some results of G. Cauchon, cf. [108], or also [66] p. 234, yields that a gr-field Δ will be an Azumaya algebra over its center if and only if $\Delta \cong \Delta_0[X,X^{-1},\varphi]$ where the automorphism φ of Δ_0 has the property that some power of it is an inner automorphism. In this case we have that $Z(\Delta) = Z(\Delta_0)^{\varphi}[\lambda X^e,(\lambda X^e)^{-1}]$, where $Z(\Delta_0)^{\varphi}$ is the fixed field for the action of φ on $Z(\Delta_0)$, λ is an appropriate unit of Δ and $e \in \mathbb{N}$ is the smallest positive integer for which there is a central element of that degree (that $e \neq 0$ follows from [108]). The conclusion of these structure results is that graded Azumaya algebras over a gr-field are of the form $M_n(\Delta_0[X,X^{-1},\varphi])(\underline{d})$, for some $n \in \mathbb{N}$, $\underline{d} \in \mathbb{Z}^n$, a skewfield Δ_0 and an automorphism φ of Δ_0 some power of which becomes an inner automorphism of Δ_0. Conversely each algebra of this form is an Azumaya algebra because of Proposition VI.3.4.

VI.3.5. Remark. If A is gr-simple gr-Artinian then A_0 is semisimple Artinian. If $\mathbb{Z}(A)$ is strongly graded then A_0 is simple Artinian (cf. [66]).

VI.3.6. Remark. Every graded finitely generated projective module over a gr-field is free, hence the graded Azumaya algebras A and B will be gr-equivalent if and only if there exist $n,m \in \mathbb{N}$, $\underline{d} \in \mathbb{Z}^n$, $\underline{e} \in \mathbb{Z}^m$ such that $M_n(A)(\underline{d}) \cong M_m(B)(\underline{e})$. Consequently, for a gr-field $k[T,T^{-1}]$ the morphism $Br^g(k[T,T^{-1}]) \to Br(k[T,T^{-1}])$ is injective.

The description of the algebras representing elements of the graded Brauer groups as well as the description of the group structure of the Brauer group depend heavily on the possibility of representing each class by a crossed product algebra over some Galois extension of the ground ring. We aim to give conditions under which certain elements of the graded Brauer group may be represented by graded crossed products. The first case we deal with is that of the graded Brauer group of a gr-field R and in this case a complete description of $Br^g(R)$ in terms of graded crossed products is possible. We need :

VI.3.7. PROPOSITION. *Let* R *be the gr-field* $k[T,T^{-1}]$ *with* $\deg T = t$. *Any graded Azumaya algebra* A *over* R *has a graded Galois splitting ring of the form* $l[T,T^{-1}] = S$ *where* l/k *is a Galois extension of* k, *with Galois group* G *say.*

PROOF. From the description of graded Azumaya algebras over gr-fields given before we retain that $A \cong M_n(\Delta_0[X,X^{-1},\varphi])(\underline{d})$ for some $n \in \mathbb{N}$, $\underline{d} \in \mathbb{Z}^n$ and some skewfield Δ_0 (see the remarks following Proposition VI. 3.4.). If l_0 is a maximal commutative subfield of Δ_0 (assuming $l_0 \neq \Delta_0$) then $A' = A \otimes_{k[T,T^{-1}]} l_0[T,T^{-1}]$ is again of the form $M_m(\Delta_0'[Y,Y^{-1},\psi])(\underline{e})$, but with $m > n$ and $\dim_{l_0} \Delta_0' < \dim_{Z(\Delta_0)} \Delta_0$. Up to a finite number of repetitions of this construction we may assume that $A' \underset{gr}{\sim} l_0[X,X^{-1},\varphi]$ where φ is some automorphism of the commutative field l_0, $l_0^{\varphi} = k$. However it is easy to check that $l_0[T,T^{-1}]$ will be a maximal commutative subring of $l_0[X,X^{-1},\varphi]$, so $l_0[T,T^{-1}]$ will be a graded splitting ring for A. It is also clear that at each step in the process the extension l_0/k may be chosen to be separable, so up to normalizing if necessary we may assume that l_0/k is a Galois extension. □

VI.3.8. THEOREM. *Let* R *be a gr-field* $k[T,T^{-1}]$ *with* $\deg T = t$. *Every graded Azumaya algebra* A *over* R *is gr-equivalent to a crossed product* $l[T,T^{-1}][u_\sigma, \sigma \in G]$ *where* l/k *is a Galois extension with Galois group* G, *the elements* u_σ, *such that* $u_\sigma\lambda = \sigma(\lambda)u_\sigma$ *for all* $\lambda \in l[T,T^{-1}] = S$, *are homogeneous and the elements* $\tau \in G$ *such that* $\deg u_\tau = 0$ *build a subgroup* G_0 *of* G *such that* G/G_0 *is* t*-torsion.*

PROOF. From the long exact sequence (Chase-Harrison-Rosenberg) :

$$1 \to H^1(G,U(S)) \to \text{Pic } R \to \text{Pic}(S)^G \to H^2(G,U(S)) \to \text{Br}(S/R) \to H'(G,\text{Pic}(S))$$

and from $\text{Pic}(R) = \text{Pic}(S) = 1$, it follows that $H^2(G,U(S)) = \text{Br}(S/R)$. By the foregoing proposition a Galois graded splitting ring $S = l[T,T^{-1}]$ for A does exist, hence we may replace A by a gr-equivalent Azumaya algebra B containing S. The latter is a crossed product algebra of the form :

$$l[T,T^{-1}][u_\sigma,\sigma \in G]$$

where $u_\sigma u_\tau = c(\sigma,\tau)u_{\sigma\tau}$, $\sigma,\tau \in G$, and $c : G \times G \to U(l[T,T^{-1}])$, $(\sigma,\tau) \mapsto c(\sigma,\tau)$, is representing an element of $H^2(G,U(S))$. If u_σ is not homogeneous, then $u_\sigma s = \sigma(s)u_\sigma$ yields $(u_\sigma)_{j_i}s = \sigma(s)(u_\sigma)_{j_i}$, where $u_\sigma = (u_\sigma)_{j_1} + \ldots + (u_\sigma)_{j_k}$, is the homogeneous decomposition of u_σ (we used that σ of degree zero). Therefore it follows that $u_\sigma^{-1}(u_\sigma)_{j_i}$ commutes with S, i.e. for an index j_i such that $(u_\sigma)_{j_i} \neq 0$ we get $(u_\sigma)_{j_i} = u_\sigma s$ for some $s \in S$. Taking parts of minimal degree in the latter equality we obtain either $(u_\sigma)_{ji}s_{min} = 0$, or else $(u_\sigma)_{ji}s_{min} = (u_\sigma)_{ji}$. The first possibility has to be excluded because $s_{min} \in S$ is invertible in the gr-field S. If we assume that $j_1 \neq j_k$ then from

$$(u_\sigma)_{ji}s_{min} = (u_\sigma)_{j_i}$$

it follows that : $(u_\sigma)_{j_k}s_{max} = 0$ (s_{min}, resp. s_{max}, denotes the homogeneous component of minimal, resp. maximal, degree in the decomposition of s). Therefore $(u_\sigma)_{j_k} = 0$ is a contradiction; consequently $j_1 = j_k$ and u_σ is a homogeneous element. Up to multiplying the u_σ by powers of T we may suppose that $|\deg u_\sigma| \leq \frac{t}{2}$ and then it is easily seen that $u_\sigma u_\tau = c(\sigma,\tau)u_{\sigma\tau}$ implies that $|\deg c(\sigma,\tau)| \leq \frac{3t}{2}$. Since $c(\sigma,\tau)$ is of the form λT^ν for some $\lambda \in k$ and $\nu \in \mathbb{Z}$, the foregoing inequality entails that

$$\deg c(\sigma,\tau) \in \{-t,0,t\}$$

for all $\sigma,\tau \in G$. We may take $u_1 = 1$. Then $u_\sigma u_{\sigma^{-1}} = c(\sigma,\sigma^{-1})$ yields : $\deg u_\sigma + \deg u_{\sigma^{-1}} \equiv 0$ modulo t. Put $G_0 = \{\sigma \in G;\ \deg u_\sigma = 0\}$. From $u_\tau u_\sigma u_{\tau^{-1}} = c(\tau,\sigma)c(\tau\sigma,\tau^{-1})u_{\tau\sigma\tau^{-1}}$, we deduce that, if $\sigma \in G_0$ and $\deg u_\tau + \deg u_{\tau^{-1}} = 0$, then $\deg u_{\tau\sigma\tau^{-1}} = 0$ while on the other hand,

$\deg u_\tau + \deg u_{\tau^{-1}} = \pm t$ implies that $\deg u_{\tau\sigma\tau^{-1}} = 0$ because $\deg u_{\tau\sigma\tau^{-1}} \equiv 0$ modulo t and $|\deg u_{\tau\sigma\tau^{-1}}| \leq \frac{t}{2}$. Consequently G_0 is a normal subgroup of G. Take $\sigma \in G - G_0$ and write $\deg u_\sigma = r$ with $0 < r \leq \frac{t}{2}$. From

$$(u_\sigma)^t = c_{\sigma,\sigma}, c_{\sigma^2,\sigma}, \ldots, c_{\sigma^{-1},\sigma} u_{\sigma^t},$$

it follows that $\deg u_{\sigma^t} \equiv 0$ modulo t hence $\deg u_{\sigma^t} = 0$ because we assumed the u_σ reduced by powers of T to the case $|\deg u_\sigma| \leq \frac{t}{2}$. So $\sigma^t \in G_0$ and thus it follows that G/G_0 is a t-torsion group. □

VI.3.9. COROLLARY. *A graded Azumaya algebra* A *over* $k[T,T^{-1}]$ *is gr-equivalent to* $\Delta_0[T,T^{-1}]$ *for some skewfield* Δ_0 *if* A *has a graded splitting ring* S *with* $\mathrm{Gal}(S/k[T,T^{-1}]) = G$ *such that* $1 = (\deg T, |G|)$.

PROOF. From $(t,|G|) = 1 = (t,|G_0|)$ it follows that any $\sigma \in G_0$ has a tth-root in G_0. So if $\tau \in G - G_0$ then $\tau^t \in G_0$ and thus $\tau^t = \gamma^t$ for some $\gamma \in G_0$ (i.e. γ may be taken to be $(\tau^t)^\alpha$ where α is such that $\alpha t + \beta|G| = 1$, $\alpha,\beta \in \mathbb{Z}$), so $\tau = \gamma$ follows. The latter contradicts $\tau \in G - G_0$. But $G = G_0$ entails that $A = A_0[T,T^{-1}]$ and A is gr-equivalent to $\Delta_0[T,T^{-1}]$ if Δ_0 is the skewfield representing A_0. □

VI.3.10. Remarks.

1. In Proposition VI.3.7. we used the fact that if a graded Azumaya A over R is gr-split by S then there is a B gr-equivalent to A such that B contains S. This is easy to prove if R is a gr-field; in general it follows from the graded version of a result of M. Auslander, O. Goldman [10].

2. We return to the problem whether every graded Azumaya algebra over a gr-field may be given by a cocycle in degree zero in the section on graded cohomology.

3. If k is a perfect field, then $\mathrm{Br}(k[T,T^{-1}])$ may be calculated as follows :

$$\begin{aligned}\mathrm{Br}(k[T,T^{-1}] &= \varinjlim_{l/k} H^2(\mathrm{Gal}(l/k), U(l[T,T^{-1}]))\\ &= \varinjlim_{l/k} H^2(\mathrm{Gal}(l/k), l^*) \oplus \varinjlim_{l/k} H^2(\mathrm{Gal}(l/k), \mathbb{Z})\\ &= \mathrm{Br}(k) \oplus (\mathrm{Gal}(\bar{k}/k))^*,\end{aligned}$$

where $\bar{k}$ is the algebraic closure of k.
If k is not perfect one may construct Azumaya algebras over $k[T]$ and also over $k[T,T^{-1}]$ which cannot be split by an extension of the form $l[T,T^{-1}]$ with l separable over k. It is a strange effect of gradations that when the Azumaya algebras considered have to be graded, then the examples (as the one given in [10]) constructed from some purely inseparable field extension are avoided.

It is known that for a Dedekind ring D the morphism $\mathrm{Br}(D) \to \mathrm{Br}(K)$ is injective, where K is the field of fractions of D, cf. [68]. If R is a gr-Dedekind ring then K^g is a gr-field hence a Dedekind ring and thus $\mathrm{Br}(K^g) \subset \mathrm{Br}(K)$. Furthermore :

VI.3.11. PROPOSITION. *Let* R *be a gr-Dedekind ring, then the canonical morphism* $\mathrm{Br}^g(R) \to \mathrm{Br}(R)$ *is a monomorphism.*

PROOF. Suppose that A and B are graded Azumaya algebras over R such that $C = A \otimes_R B^0 \cong \mathrm{End}_R(P)$ for some R-progenerator P. With the ususal gradation C is a graded Azumaya algebra over R, hence its total graded ring of quotients $Q^g(C)$ is an Azumaya algebra over $K^g = k[T,T^{-1}]$ (where $Q^g(C) = C \otimes_R K^g$). By Remark VI3.6. and the remarks preceding this proposition it is clear that the morphisms in the following row are injective morphisms :

$$\mathrm{Br}^g(k[T,T^{-1}]) \to \mathrm{Br}(k[T,T^{-1}]) \to \mathrm{Br}(k(T)).$$

So $Q^g(C) \cong \mathrm{END}(V_n)$ for some graded gr-free $k[T,T^{-1}]$-module V_n, of rank n say; in other words $C \otimes_R k[T,T^{-1}] \cong \mathrm{END}(V_n)$ as graded R-algebras. Up to shifting V_n if necessary we may assume that $(V_n)_0 \neq 0$. Pick any $v \in (V_n)_0$ and define a map

$$\psi : C \underset{R}{\otimes} k[T,T^{-1}] = \mathrm{END}(V_n) \to V_n,\ f \mapsto f(v).$$

Now $E = \psi(C)$ is a graded finitely generated and torsion free R-module, so by Corollary VI.1.14, E is projective. We obtain an embedding of degree zero $C \to \mathrm{HOM}_R(E,E) = \mathcal{E}$, because E is graded faithful as an R-module and as a C-module. The commutator theorem, cf. [33], Theorem 4.3. implies that $\mathcal{E} = C \underset{R}{\otimes} \mathcal{E}^{(C)}$ where $\mathcal{E}^{(C)}$ consists of the elements in $\mathcal{E}$ which commute with all elements of C (we identify C with its image in $\mathcal{E}$). Counting ranks yields $\mathcal{E} = C$ i.e. $A \underset{R}{\otimes} B^0 = \mathrm{END}(P')$ where P' is a graded R-progenerator and thus A is gr-equivalent to B. □

VI.3.12. <u>COROLLARY</u>. *We have the following commutative diagram of injective group homomorphisms :*

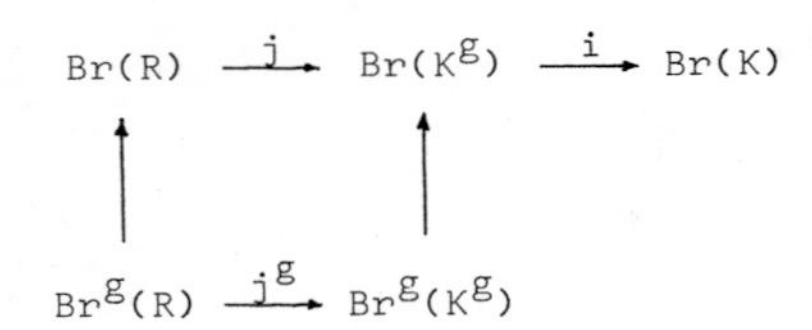

<u>PROOF</u>. Commutativity is easily checked. The vertical morphisms are injective because of Proposition VI.3.11. Since K^g is a Dedekind ring, i is injective. Since R is a Noetherian integrally closed domain of dimension 2, ij is injective and thus j is injective. Now injectivity of j^g follows from the commutativity of the diagram (or from the easy graded version of the proof of the fact that the Brauer group of a Dedekind domain embeds in the Brauer group of its field of fractions.) □

VI.3.13. <u>COROLLARY</u>. *Let* A *be a graded Azumaya algebra over a gr-Dedekind ring* R.

1. *Let* S *be a graded commutative integrally closed domain containing* R *as a graded subring and such that* S *is a finitely generated* R*-module. If* <u>S</u> *splits* <u>A</u> *then* S *is a gr-splitting ring for* A.

2. *Let* S *be a commutative graded separable extension of* R. *If* S *splits* A *then* S *is a gr-splitting ring for* A.

PROOF. 1. It is clear that S is the integral closure of R in the gr-field of fractions $L^g = Q^g(S)$ of S. By Corollary VI.1.10.1., S is again a gr-Dedekind ring. Hence $A \otimes_R S = \text{End}_S(P')$ for some S-progenerator P' entails that $A \otimes_R S \cong \text{END}_S(P)$, as graded R-algebras, for some graded R-progenerator P.

2. Any finitely generated torsion free graded S-module M is by restriction of scalars a finitely generated torsion free graded R-module, i.e. projective as an R-module. The separability of the extension S/R entails that M is also projective as an S-module. Now we may use this in the proof of Proposition VI.3.11. with R replaced by S. Consequently, the argumentation of that proof leads to the fact that $A \otimes_R S \sim 1$ in Br(S) implies $A \otimes_R S \underset{gr}{\sim} 1$ in $Br^g(S)$. □

We will come back to further explicit calculations in Section VI.5., where the relations between $Br^g(-)$ and Br(-) will be investigated further.

Now we look back at Section III.3. Since R-gr is a Grothendieck category it is possible to develop the theory of graded localization via graded torsion theories, e.g. [66] for a non-commutative graded theory. Then one can use graded kernel functors K^g and filters having a basis of graded ideals in order to develop a theory of relative graded Brauer groups $Br^g(R,K^g)$ parallel to the theory expounded in Section III.3. Thinking further in the same vein one may also develop some techniques for $Pic^g(R,K^g)$, $Pic_g(R,K^g)$ adapting the results of Chapter II to the graded situation. Although this routine exercise may be fun we do not need to go into the details here. Indeed we have been prompted to study relative invariants of rings mostly by geometrical considerations and this is still true for the graded theory. It will turn out that the study of Brauer groups of projective schemes or varieties depends less on

$Br^g(-)$ than one might hope for, as a matter of fact it is a graded relative Brauer group of the homogeneous coordinate ring which relates directly to the Brauer group of the projective variety. However we will only need the relative graded Brauer group with respect to the graded kernel functor corresponding to graded minimal nonzero prime ideals, so this explains why we are here only interested in the graded version of Yuan's group $\beta(R)$, cf. [103] and Chapter V, Section V.2. Yuan's results may be generalized to arbitrary Krull domains, cf. M. Orzech [67] and the same is true for the results we will mention about graded Noetherian integrally closed domains. Nevertheless we restrict to the strict necessary here; in the set of exercises at the end of the book the reader may find some hints towards a more general treatment of the material presented.

In the sequel of this section, R will be a $\mathbb{Z}$-graded Noetherian integrally closed domain. Recall that $X = \mathrm{Spec}(R)$, $X^{(1)} = \{P \in X;\ \mathrm{ht}\ P = 1\}$; we also introduce $X_g = \{P \in X;\ P \text{ is graded}\}$, $X_g^{(1)} = \{P \in X^{(1)};\ P \text{ is graded}\}$. Note that for a $P \in X^{(1)}$ either $P_g = 0$ or $P_g = P$ and in the latter case P is graded. We write K^g for the gr-field of homogeneous fractions of R and K for the field of fractions of R of K^g. If M is a graded R-module then we have canonical localization morphisms

$$M \to M \underset{R}{\otimes} K^g \to M \underset{R}{\otimes} K.$$

These maps are monomorphisms if M is torsion free. For any family of multiplicative systems $\{S_i;\ i \in J\}$ in R we write $\bigcap_{i\in J} S_i^{-1}M$ for the intersection in $M \underset{R}{\otimes} K$ of the images of $S_i^{-1}M$ in $M \underset{R}{\otimes} K$. If each S_i, $i \in J$, consists of homogeneous elements then each $S_i^{-1}M$ maps into $M \underset{R}{\otimes} K^g$ and then the intersection $\bigcap_{i\in J} S_i^{-1}M$ may be taken in $M \underset{R}{\otimes} K^g$. If $M \in R\text{-gr}$ then M^* now stands for $\mathrm{HOM}_R(M,R)$ so if M is not finitely generated one should be careful when handling M^* because the ungraded dual $(\underline{M})^*$ is different (in general) from $\underline{M}^*$.

An $M \in R\text{-gr}$ is said to be <u>gr-reflexive</u> if the canonical map $c : M \to M^{**}$ is an isomorphism in R-gr (note that c maps $m \in M$ to

$\hat{m} \in HOM_R(M^*,R)$ which is given by $\hat{m}(\varphi) = \varphi(m)$ for $\varphi : M \to R$). If $M \in R$-gr is finitely generated then M is gr-reflexive if and only if $\underline{M}$ is reflexive in R-mod.

We aim to give a criterion for gr-reflexivity in terms of graded localizations at graded minimal prime ideals. The set

$$(X^{(1)})_g = \{P_g;\ P \in X^{(1)}\}$$

is $X_g^{(1)}$ plus possibly the zero ideal.

VI.3.14. LEMMA. *If* R *is any graded Noetherian commutative ring then* $(X^{(1)})_g \neq \{0\}$ *unless* R *is a gr-field.*

PROOF. This is essentially a consequence of Krull's principal ideal theorem. If R is not a gr-field then there is an $x \neq 0$ in h(R) which is not invertible in R. The principal ideal theorem then yields that any prime ideal P of R minimal over Ru is in $X^{(1)}$. However : $Ru \subset P_g \subset P$ implies $P_g = P$, hence $P \in (X^{(1)})_g$. □

VI.3.15. LEMMA. *If* M *is a finitely generated torsion free* R*-module, where* R *is an integrally closed Noetherian domain, then*

$$M^{**} = \cap\{Q_{R-P}(M);\ P \in X^{(1)}\}.$$

PROOF. Cf. [36]. □

VI.3.16. COROLLARY. *Let* R *be a graded Noetherian integrally closed domain and suppose that* $M \in R$-gr *is finitely generated and torsion free. Then,* $M^{**} = \cap\{Q^g_{R-P}(M);\ P \in X_g^{(1)}\}$.

PROOF. Using the techniques of Section V.2. and Section III.1. in particular Proposition V.2.3., (or actually its proof) we have that $M^{**} \subset Q_{R-P}(M)$ for all $P \in X^{(1)}$. Since M^{**} is graded $M^{**} \subset N$ where $N = \cap\{Q^g_{R-P}(M);\ P \in X_g^{(1)}\}$. Conversely, if $z \in h(N)$ then we may find a graded ideal I_P for each $P \in X_g^{(1)}$ such that $I_P z \subset N$. Hence $(\sum_{P \in X^{(1)}} I_P)z \subset M$. If $P' \in X^{(1)}$ is not graded then $I = \sum_{P \in X_g^{(1)}} I_P \not\subset P'$ because otherwise $I \subset P'_g \underset{\neq}{\subset} P'$ entails $I = 0$, a contradiction. Therefore $I \not\subset P$ for every $P \in X^{(1)}$. Then $Iz \subset M$ implies that

$$z \in \cap\{Q_{R-P}(M);\ P \in X^{(1)}\} = M^{**}$$

and thus $N = M^{**}$ as desired. □

VI.3.17. COROLLARY. *With assumptions as in VI.3.16., the following statements are equivalent :*

1. $M \in R\text{-gr}$ *is gr-reflexive;*
2. $M = \cap\{Q^g_{R-P}(M);\ P \in X_g^{(1)}\}$.

PROOF. Clear. □

An $M \in R\text{-gr}$ is said to be gr-quasireflexive if

$$M = \cap\{Q^g_{R-P}(M);\ P \in X_g^{(1)}\}$$

i.e. if $M = Q^g_K(M)$ where $K = \inf\{K^g_{R-P};\ P \in X_g^{(1)}\}$. It is clear that for torsion free finitely generated M the notions of gr-quasireflexivity and gr-reflexivity coincide.

VI.3.18. Examples.

1. Let R be a graded Noetherian integrally closed domain and let $M \in R\text{-gr}$ be a gr-flat module, then M is gr-quasireflexive.

2. An integral extension S of R which is again Noetherian and integrally closed is gr-quasireflexive as an R-module.

PROOF. 1. If M is gr-flat then $\underline{M}$ is flat in R-mod. From Chapter II we recall that for any idempotent kernel functor κ on R-mod, and for all $\underline{N} \in R\text{-mod}$, we have $Q_\kappa(\underline{N} \otimes_R \underline{M}) = Q_\kappa(\underline{N}) \otimes_R \underline{M}$. If we apply this to κ defined above with $N = R$ then we obtain $\underline{M} = Q_\kappa(\underline{R}) \otimes_R \underline{M} = Q_\kappa(\underline{R} \otimes_R \underline{M}) = Q_\kappa(\underline{M})$.

2. The "going down" property for the extension S of R entails that $P \cap R \in (X^{(1)})_g(R)$ if $P \in X_g^{(1)}(S)$ and then

$$S = \cap\{Q^g_{R-P}(S);\ P \in X_g^{(1)}(S)\} = \cap\{Q^g_{R-P}(S);\ p \in X_g^{(1)}(R)\}$$

VI.3.19. PROPOSITION. *Let* R *be as before and strongly graded then there is a bijective correspondence between*

1°. *gr-quasireflexive* $M \in R\text{-gr}$;

2°. *quasireflexive* $N \in R_0\text{-mod}$.

PROOF. The results about strongly graded rings, cf. [66], entail that R is Noetherian and integrally closed if and only if R_0 is Noetherian and integrally closed. Moreover graded prime ideals of R correspond bijectively to prime ideals of R_0 under the correspondence $P \twoheadrightarrow P \cap R_0$ (while $P = R(P \cap R_0)$) if P is a graded prime ideal of R. By Lemma VI.1.2.2. $Q^g_{R-P} = Q_{R_0-p}$ where $p = P \cap R_0$. By Lemma VI.1.2.1. $M = M_0 \underset{R_0}{\otimes} R$, and $Q^g_{R-P}(M) = Q_{R_0-p}(M_0) \underset{R_0}{\otimes} R$. Taking intersections over $p \in X^{(1)}(R_0)$ i.e. over $P \in X_g^{(1)}(R)$ and using flatness of R as an R_0-module, the statement becomes evident. □

The modified tensor product of graded R-modules M and N may now be defined by : $M \underset{g}{\perp} N = Q^g_\kappa(M \underset{R}{\otimes} N)$, where κ is defined as before. This modified tensor product $\underset{g}{\perp}$ satisfies the usual properties also satisfied by $\perp$, introduced in V.2.

VI.3.20. PROPOSITION. *Let* $E \in R\text{-gr}$ *be finitely generated and reflexive, then* $E^* \underset{g}{\perp} E = END_R(E)$.

PROOF. Since $\underline{E}$ is finitely generated and reflexive in R-mod we have that $E = Q^g_\kappa(E)$ where $\kappa = \inf\{\kappa^g_{R-P};\ P \in X_g^{(1)}\}$. Since E is finitely generated : $END_R(E) = End_R(E)$. From Yuan [103], it follows that $E^* \perp E = End_R(E)$ therefore the statement follows if one checks that $M \underset{g}{\perp} N$ for $M,N \in R\text{-gr}$ has $\underline{M} \perp \underline{N}$ for the underlying R-module. Another proof is obtained by local argumentation i.e. for all $P \in X_g^{(1)}$ we have

$Q^g_{R-P}(END_R(E)) = END_R(Q^g_{R-P}(E))$ where $Q^g_{R-P}(E)$ is a graded projective $Q^g_{R-P}(R)$-module; hence $Q^g_{R-P}(E^*) \underset{Q^g_{R-P}(R)}{\otimes} Q^g_{R-P}(E) \cong END_{Q^g_{R-P}(R)}(Q^g_{R-P}(E))$. The proof is finished if one verifies that the global graded morphism $E^* \underset{R}{\otimes} E \to END_R(E)$ inducing the forementioned "local" graded isomorphisms has degree zero. □

VI.3.21. <u>COROLLARY</u>. *If* M *is a graded finitely generated reflexive* R-*module of rank one (i.e.* $M \underset{R}{\otimes} K^g \cong K^g$ *in* R-gr) *then* $M^* \underset{g}{\perp} M = R$. □

From the foregoing it follows that the isomorphism classes of gr-reflexive finitely generated R-modules of rank one form a group with multiplication induced by the modified tensor product $\underset{g}{\perp}$. This group will be denoted by $\Gamma_g(R)$; See Section V.2. for the definition of $\Gamma(R)$.

VI.3.22. <u>PROPOSITION</u>. *If* R *is a graded Noetherian integrally closed domain then* $\Gamma^g(R) = \Gamma(R)$.

<u>PROOF</u>. There is an exact sequence $U(K) \to D(R) \to \Gamma(R) \to 0$, cf. [36]. Let us check that we also have an exact sequence :

$$U(K^g) \xrightarrow{\nu} D_g(R) \xrightarrow{\eta} \Gamma_g(R) \longrightarrow 0.$$

If this is the case then the commutativity of the following diagram (which follows from the definition of ν and η given below) finishes the proof :

$$\begin{array}{ccccccc} 0 & & 0 & & & & \\ \downarrow & & \downarrow & & & & \\ U(K^g) & \longrightarrow & D_g(R) & \longrightarrow & \Gamma_g(R) & \longrightarrow & 0 \\ \downarrow & & \downarrow & & \downarrow & & \\ U(K) & \longrightarrow & D(R) & \longrightarrow & \Gamma(R) & \longrightarrow & 0 \end{array}$$

We define ν by $\nu(q) = \Sigma\, v_P(q).[P]$ in the divisor group generated by symbols $[P]$ for $P \in X^{(1)}_g$, and where $v_P(q)$ is the valuation of q with respect to the graded valuation associated to the gr-valuation ring $Q^g_{R-P}(R)$

i.e. $qQ^g_{R-P} = P^{v_P(q)}Q^g_{R-P}(R)$. We define η by $\eta([P]) = P$. Exactness of the sequence as well as commutativity of the diagram is easily checked. □

In the definition of $\Gamma_g(R)$ the important feature is that $M^* \perp_g M = \mathrm{End}(M) = R$, so it will be natural to look at graded R-algebras A with the property $A^* \perp_g A = \mathrm{END}(A) = \mathrm{End}(A)$ and such that A is a finitely generated reflexive R-module. Such algebras are called graded reflexive Azumaya algebras. Two such algebras, A and B say, are said to be graded relative similar if we may find graded finitely generated reflexive R-modules, E and F say, such that $A \perp_g \mathrm{END}(E) \hat{=} B \perp_g \mathrm{END}(F)$ as graded rings. One easily checks the following properties :

VI.3.23. PROPOSITION. *Let* R *be as before.*

1. *If* E,F *are graded finitely generated reflexive* R*-modules, then* $\mathrm{END}(E) \perp_g \mathrm{END}(F) \cong \mathrm{END}(E \perp_g F)$, *as graded rings.*
2. *If* A *and* B *are graded reflexive Azumaya algebras over* R, *then so is* $A \perp_g B$.
3. *Graded relative similarity is an equivalence relation.*
4. *If* A *is a graded reflexive algebra over* R *then the following statements are equivalent :*
   a) A *is a graded reflexive Azumaya algebra over* R;
   b) for all $P \in X^{(1)}_g(R)$, $Q^g_{R-P}(A)$ *is a graded Azumaya algebra over* R.
5. *If* E *is a finitely generated graded reflexive* R*-module then* END(E) *is a graded reflexive Azumaya algebra over* R.

It is clear that the statements about "reflexive" objects are just a special case of the relative theory one may introduce by considering graded kernel functors κ^g by taking for κ^g exactly $\inf\{\kappa^g_{R-P};\ P \in X^{(1)}_g(R)\}$. A straightforward combination of graded and relative techniques yields the following (with assumptions as before) :

VI.3.24. THEOREM. *The set of similarity classes of graded reflexive Azumaya algebras over* R *forms a group* $\beta^g(R)$ *with respect to the modified tensorproduct* $\underset{g}{\perp}$. *If* R *is strongly graded then* $\beta^g(R) = \beta(R_0)$, *where* $\beta(R_0)$ *is the algebra class group introduced by Yuan in* [103] *(see also Section V.2.).*

VI.3.25. THEOREM.

1. *For all* $P \in X_g^{(1)}$ *we have a monomorphism* $\beta^g(R) \hookrightarrow Br^g(Q^g_{R-P}(R))$.
2. $\beta^g(R) = \cap\{Br^g(Q^g_{R-P}(R)),\ P \in X_g^{(1)}\} = \cap\{Br(R_{(P)},\ P \in X_g^{(1)}\}$.
3. *If* R *is a gr-Dedekind domain then :* $\beta^g(R) = Br^g(R)$.

PROOF. 1. Let A be a graded reflexive Azumaya algebra over R such that $Q^g_{R-P}(A)$ is trivial in $Br^g(Q^g_{R-P}(R))$, for some $P \in X_g^{(1)}$. For each $P \in X_g^{(1)}$, $Q^g_{R-P}(A)$ is a graded Azumaya algebra over $Q^g_{R-P}(R)$, hence a maximal $Q^g_{R-P}(R)$-order in $A \underset{R}{\otimes} K$. Mimicking the proof of a similar result obtained by M. Auslander, O. Goldman in [10], we deduce from the above that $A = END_R(E)$ for some finitely generated graded reflexive R-module E i.e. [A] is trivial in $\beta^g(R)$.

2. Since each $Q^g_{R-P}(R)$ is a gr-Dedekind ring,

$$B = \cap\{Br^g(Q^g_{R-P}(R);\ P \in X_g^{(1)}\} \subset Br(K^g).$$

Consider a gr-field Δ in $Br(K^g)$ such that for each $P \in X_g^{(1)}(R)$ there exists a graded Azumaya algebra A(P) over $Q^g_{R-P}(R)$ such that $A(P) \underset{Q^g_{R-P}(R)}{\otimes} K^g$ is gr-equivalent to Δ.

We have to show that there is a graded reflexive Azumaya algebra A over R such that $Q^g_{R-P}(A) \sim A(P)$ for all $P \in X_g^{(1)}(R)$. Let A_1 be any gr-maximal R-order in Δ (cf. [113] for some detail on gr-maximal orders) i.e. A_1 is a maximal order and a graded R-algebra (cf. [115]). Then $A_1 = A_1^{**}$ by maximality and furthermore $Q^g_{R-P}(A)$ is a maximal (graded) order over $Q^g_{R-P}(R)$ in Δ. Since $Q^g_{R-P}(R)$ is a gr-principal ideal ring it follows that A(P) and $Q^g_{R-P}(A_1)$ are gr-equivalent for all $P \in X_g^{(1)}(R)$. Consequently, for all $P \in X_g^{(1)}(R)$, $Q^g_{R-P}(A_1)$ is a graded Azumaya algebra, so by Proposition VI.3.23.4., it follows that A_1 is a graded reflexive Azumaya algebra over R. Take $A = A_1$; it has the desired property.

3. In this case R is a regular ring of global dimension less than or

equal to two. Consequently the finitely generated (graded) reflexive R-modules will be finitely generated (graded) projective R-modules and therefore the notions of $\beta^g(R)$ and $Br^g(R)$ coincide. □

VI.4. Mayer-Vietoris Sequences for Graded Brauer Groups.

The exact sequences of Mayer-Vietoris type may be used to calculate the graded Brauer groups. In the geometric context these sequences appear when we calcuate Brauer groups of varieties with singularities. Let us first (re)introduce some of the abstract machinery.

Throughout this section R is a commutative graded ring with unit. Let $FP_g(R)$ be the category of graded faithfully projective R-modules, i.e. faithful finitely generated graded projective R-modules, with graded R-isomorphisms of degree zero for the morphisms. The forgetful functor $\underline{\ }$: R-gr → R-mod, restricts to $FP_g(R) \to FP(R)$, because a graded projective R-module is a projective R-module in the ungraded sense. The product in the category $FP_g(R)$ is the one induced by the tensor product $\otimes_R$ with the usual gradation on it.

Let $AZ_g(R)$ consist of the graded Azumaya algebras over R, let the morphisms in this category be given by the graded R-algebra morphisms of degree zero. Then the tensor product $\otimes_R$ induces a product in this category as well.

The category $\underline{Pic}(R)$ introduced in Chapter II has a graded analogue $\underline{Pic}_g(R)$, consisting of the graded invertible R-modules (i.e, graded R-modules P such that there is a graded R-module Q such that $P \otimes_R Q \cong R$ as graded R-modules) and graded R-isomorphisms of degree zero. The product in $\underline{Pic}_g(R)$ is again induced by the tensor product $\otimes_R$. Obviously, for all $n \in \mathbb{Z}$, $R(n) \in \underline{Pic}_g(R)$, since $R(n) \otimes_R R(-n) \cong R$ in R-gr.
Note that in defining FP_g, AZ_g and $\underline{Pic}_g$ we utilized graded morphisms which are of degree zero; in a formally similar way one may introduce

FP^g, AZ^g and $\underline{Pic}^g$ by using morphisms not necessarily of degree zero for the category morphisms. We have chosen the first set up because e.g. $K_0\ \underline{Pic}_g = Pic_g$. The terminology and concepts of Chapter II will be used freely throughout this section (i.e. K_0, K_1, E-surjective, etc...).

VI.4.1. LEMMA. *For any* $P \in$ R-gr *the following statements are equivalent :*

1. *There exist* $Q \in$ R-gr *and* $n \in \mathbb{N}$ *such that for some* $(p_1,\ldots,p_n) \in \mathbb{Z}^n$ *we have that* $P \otimes_R Q$ *and* $S = \bigoplus_{i=1}^n R(p_i)$ *are isomorphic in* R-gr.
2. P *is a graded faithfully projective* R-*module.*

PROOF. 1. ⇒ 2. (Following the proof of the similar ungraded statement in [12]). Assume $P \otimes_R Q = \bigoplus_{i=1}^n R(p_i)$, then $\varepsilon_i = (0,\ldots,1,0,\ldots,0)$ with 1 in the ith place is in S_{-p_i} and it corresponds to some element of the form $\sum_j x_j \otimes y_j$ in $(P \otimes_R Q)_{-p_i}$.
Fix a finite family $\{x_\alpha \otimes y_\alpha;\ \alpha \in J\}$ in $P \otimes_R Q$ such that this family is a homogeneous set of generators for the graded R-module $P \otimes_R Q$. Now we define a morphism in R-gr (of degree zero !) $\mu : \bigoplus_\alpha R(-d_\alpha) \to P$, $\varepsilon_\alpha \mapsto x_\alpha$, where d_α is the degree of x_α in P. Tensoring by Q yields a surjective graded morphism of degree zero $\mu \otimes_R Q : \bigoplus_\alpha R(-d_\alpha) \otimes_R Q \to P \otimes_R Q$, and by the assumptions of 1. this is a split epimorphism in R-gr. Furthermore, $\mu \otimes_R Q \otimes_R P$ is then also a split epimorphism in R-gr. But

$$\mu \otimes_R Q \otimes_R P = \mu \otimes \left(\bigoplus_{i=1}^n R(p_i)\right) = \bigoplus_{i=1}^n (\mu \otimes_R R(p_i))$$

is just a direct sum of "shifted copies" of μ, therefore each $\mu \otimes_R R(p_i)$ and consequently μ itself is a split epimorphism in R-gr. This establishes that P is a graded direct summand of a gr-free module of finite rank i.e. P is finitely generated and projective. The same argument holds for Q. Finally, both P and Q are graded faithfully projective because $P \otimes_R Q$ is.

2. ⇒ 1. Recall Lemma A.6. of [9]; this states that if M is a finitely generated projective R-module and $S = Hom_R(M,M)$, then for any left S-module N we have that the canonical map :

$$\rho : \mathrm{Hom}_S(M,N) \underset{R}{\otimes} M \to N,$$

will be an isomorphism whenever the trace map $\mathrm{Hom}_R(M,R) \underset{S}{\otimes} M \to R$ is surjective. If we now take $M = P$ as in 2. then we first present P as a direct summand of a gr-free R-module $\bigoplus_{i=1}^{n} R(q_i)$, $(q_1,\ldots,q_n) \in \mathbb{Z}^n$, i.e. we find a $Q \in R$-gr such that $P \oplus Q = \bigoplus_{i=1}^{n} R(q_i)$. Then $\mathrm{HOM}_R(P,P)$ is a direct summand of $\mathrm{HOM}_R(\bigoplus_{i=1}^{n} R(q_i), \bigoplus_{i=1}^{n} R(q_i))$ with complement :

$$\mathrm{HOM}_R(P,Q) \oplus \mathrm{HOM}_R(Q,P) \oplus \mathrm{HOM}_R(Q,Q).$$

It is thus evident that $\bigoplus_{i=1}^{n} R(q_i)$ may be viewed as an $S = \mathrm{END}_R(P)$-module and in this structure it will be a graded S-module (both S and $\mathrm{END}_R(\bigoplus_{i=1}^{n} R(q_i))$ are considered with their natural gradations of course). That the trace map $\mathrm{HOM}_R(P,R) = \mathrm{Hom}_R(P,R) \to R$ is surjective may be verified locally by localizing at maximal ideals. Therefore we may apply the general lemma quoted above and deduce from this the fact that ρ,

$$\rho : \mathrm{HOM}_S(P, \bigoplus_{i=1}^{n} R(q_i)) \underset{R}{\otimes} P \to \bigoplus_{i=1}^{n} R(q_i),$$

is an isomorphism of R-modules. If the left hand side is endowed with its usual gradation then ρ is clearly a graded morphism of degree zero and the assertion follows by taking

$$Q = \mathrm{HOM}_S(P, \bigoplus_{i=1}^{n} R(q_i)) = \mathrm{Hom}_S(P, \bigoplus_{i=1}^{n} R(q_i)). \quad \square$$

From the definitions it is clear that two graded Azumaya algebras A and B represent the same element in $\mathrm{Br}^g(R)$ if the graded isomorphism class of $A \underset{R}{\otimes} B$ contains an $\mathrm{END}_R(P)$ for some $P \in \mathrm{FP}_g(R)$. The functor $\mathrm{END}_R : \mathrm{FP}_g(R) \to \mathrm{AZ}_g(R)$ is product preserving, hence it induces a group homomorphism $K_0\ \mathrm{END}_R : K_0\mathrm{FP}_g(R) \to K_0\mathrm{AZ}_g(R)$, (see Chapter II for some definitions, cf. also [60]). The definition of $\mathrm{AZ}_g(R)$ now entails that $\mathrm{Br}^g(R)$ is just Coker $K_0\ \mathrm{END}_R$. We also have $K_0\ \underline{\mathrm{Pic}}_g(R) = \mathrm{Pic}_g(R)$. Let $t\ \mathrm{Pic}_g(R)$ be the torsion part of the group $\mathrm{Pic}_g(R)$.

VI.4.2. <u>PROPOSITION</u>. *For any $\mathbb{Z}$-graded commutative ring* R, *the following sequence of group homomorphism is exact :*

$$0 \to t\ \mathrm{Pic}_g(R) \to \mathrm{Pic}_g(R) \to K_0 FP_g(R) \to K_0 AZ_g(R) \to Br^g(R) \to 0.$$

PROOF. Exactness of $K_0FP_g(R) \to K_0AZ_g(R) \to Br^g(R) \to 0$ is a consequence of the definitions in particular of the definition of the functor K_0 END_R. Since $\mathrm{Pic}_g(R)$ is K_0 $\underline{\mathrm{Pic}}_g(R)$ exactness at $K_0KFP_g(R)$ is formally the same as the similar ungraded statement, actually if P is in $\underline{\mathrm{Pic}}_g(R)$ then $END_R(P) \cong R$ and this property may be used to characterize the image of K_0 $\underline{\mathrm{Pic}}_g(R)$ in $K_0FP_g(R)$. Let us now prove exactness at $\mathrm{Pic}_g(R)$! First take $[I] \in t\ \mathrm{Pic}_g(R)$, i.e. for some $p \in \mathbb{Z}$, $n \in \mathbb{N}$, $I^{\otimes n} \cong R(p)$ in R-gr. Put $E = R \oplus I(-d) \oplus I(-d)^{\otimes 2} \oplus \ldots \oplus I(-d)^{\otimes n-1}$ for some $d \in \mathbb{N}$. Then $E \underset{R}{\otimes} I$ will be isomorphic to E(d) in R-gr if we have that $I^{\otimes n}(-nd) = R(p-nd) = R$ i.e. when $p = nd$. Such a d is obtained by localizing homogeneously at a gr-maximal ideal ω of R, indeed we obtain :

$$Q^g_{R-\omega}(I)^{\otimes n} = Q^g_{R-\omega}(I^{\otimes n}) = Q^g_{R-\omega}(R(p)) = Q^g_{R-\omega}(R)(p)$$

on one hand. On the other hand $Q^g_{R-\omega}(I)$ is a gr-free $Q^g_{R-\omega}(R)$-module because it is a finitely generated graded projective module over the gr-local ring $Q^g_{R-\omega}(R)$. Put $Q^g_{R-\omega}(I) \cong Q^g_{R-\omega}(R)(d)$ for some $d \in \mathbb{Z}$. Then we calculate :

$$\begin{aligned} Q^g_{R-\omega}(R)(nd) &= (Q^g_{R-\omega}(R)(d))^{\otimes n} = Q^g_{R-\omega}(I)^{\otimes n} \\ &= Q^g_{R-\omega}(I^{\otimes n}) = Q^g_{R-\omega}(R)(p) \end{aligned}$$

yielding $p = nd$.

The graded R-module E constructed above is obviously a faithfully projective R-module so there exists $F \in$ R-gr such that $E \underset{R}{\otimes} F \cong \overset{r}{\underset{i=1}{\oplus}} R(s_i)$ in R-gr, for some $r \in \mathbb{N}$, $(s_1,\ldots,s_r) \in \mathbb{Z}^n$. From $E \underset{R}{\otimes} I \cong E(t)$ in R-gr we derive (by tensoring both sides with F) :

$$\begin{aligned} \overset{r}{\underset{i=1}{\oplus}} I(s_i) &= I \underset{R}{\otimes} (\overset{r}{\underset{i=1}{\oplus}} R(s_i)) = E \underset{R}{\otimes} F(d) = \overset{r}{\underset{i=1}{\oplus}} R(s_i+d) \\ &= R(d) \otimes (\overset{r}{\underset{i=1}{\oplus}} R(s_i)). \end{aligned}$$

The relations imposed in $K_0FP^g(R)$ allow to express the foregoing as :

$[I] = [R(d)] = [R] = 0$. The converse implication, i.e. if $[I] \in \mathrm{Pic}_g(R)$

vanishes in $K_0 FP_g(R)$ then $[I] \in t\ Pic_g(R)$ may be deduced in exactly the same way. □

Since the objects of $\underline{Pic}_g(R)$ have constant (graded) rank one it is possible to improve on part of the exact sequence mentioned above. To this aim we introduce the category $CRP_g(R)$ of graded finitely generated projective R-modules of constant (graded) rank and with graded R-isomorphisms for the morphisms and $\underset{R}{\otimes}$ inducing the product. Let $CR_{0g}(R)$ denote $K_0\ CRP_g(R)$ and from the foregoing it is clear that we have an exact sequence : $0 \to t\ Pic_g(R) \to Pic_g(R) \to CR_{0g}(R)$, while $CR_{0g}(R) \to K_0 FP_g(R)$ is an inclusion. The extension of this sequence to a long exact sequence as before is obviously linked to connectedness properties of $Spec_g(R)$.

VI.4.3. LEMMA. *If* R *is a gr-semilocal ring then* $Pic^g(R) = 0$.

PROOF. Finitely generated graded projective R-modules are gr-free in this case, cf. [66], so the proof is easy. □

Note added in proof.

Lemma VI.4.3. is not true in full generality as stated; the reference to [66] only applies to the gr-local case. The Lemma does hold in either of the following cases:

1° R is positively graded, e.g. for the geometrical applications.

2° R is strongly graded, e.g. for the applications to crossed product theorms over arithmetically graded rings.

3° $J^g(R) = J(R)$, e.g. if R is Artinian.

4° R is gr-local.

The reader may verify that Theorem VI.4.6. may be easily adapted to this new situation while Theorem VI.4.7. is not affected at all. By 1 or 3 the results derived on p. 175 remain unaffected too. There do exist examples of gr-semilocal gr-Dedekind rings R with $Pic^g(R) \neq o$. Let R_o be a Dedekind ring with exactly two maximal ideals M_1, M_2.

Put : $S = \ldots M_1M_2X^{-2} + M_1M_2X^{-1} + R_o + R_oX + M_1^{-1}M_2^{-1}X^2 + M_1^{-1}M_2^{-1}X^3 + \ldots$

Consider the graded ideal I,

$$I = \ldots M_1^2M_2^2X^{-1} + M_1M_2^2 + M_1M_2X + M_2X^2 + RX^3 + M_1^{-1}X^4 + \ldots$$
$$= SM_1 \cap SM_2X.$$

Clearly I is graded projective of rank 1 and not free.

The proof of Lemma VI.4.3. in each of the good cases mentioned above may be viewed as an easy exercise

1 by descent of projective graded modules, 2 by the equivalence of R_o-mod and R-gr., 3 by lifting of freeness modulo the Jacobson radical,
4 cf. [66] or a method similar to 3 using now that $R/J^g(R)$ is a graded field.

Concerning gr-semilocal rings it is also interesting to note the following (due to S. Caenepeel, [17]) :

VI.4.4. Remark. If the $\mathbb{Z}$-graded commutative ring R is gr-semilocal then the idempotent elements of R are necessarily homogeneous of degree zero.

We now turn to the application of fibred products and Milnor's theorem to certain functors related to the categories of graded objects introduced in the first part of this section. First we consider the graded fibre product of graded rings, i.e. we have a commutative diagram of graded morphisms of degree zero :

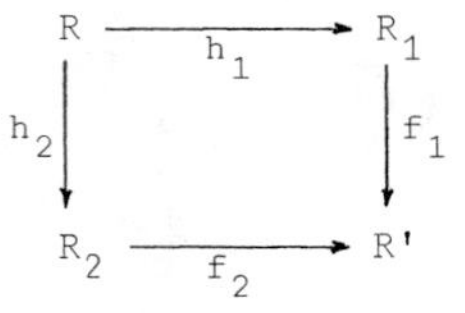

$$(*)\qquad \begin{array}{ccc} R & \xrightarrow{h_1} & R_1 \\ {\scriptstyle h_2}\downarrow & & \downarrow{\scriptstyle f_1} \\ R_2 & \xrightarrow{f_2} & R' \end{array}$$

such that R is graded isomorphic in degree zero to

$$\{(r_1,r_2) \in R_1 \times R_2; f_1(r_1) = f_2(r_2)\},$$

where $R_1 \times R_2$ may be graded by putting $(R_1 \times R_2)_n = (R_1)_n \times (R_2)_n$. Let $C^g(-)$ denote any one of the categories introduced before. The diagram of graded morphisms (*) yields a diagram of "extension of scalars" functors :

$$\begin{array}{ccc} \mathcal{C}^g(R) & \xrightarrow{H_1} & \mathcal{C}^g(R_1) \\ {\scriptstyle H_2}\downarrow & & \downarrow{\scriptstyle F_1} \\ \mathcal{C}^g(R_2) & \xrightarrow{F_2} & \mathcal{C}^g(R') \end{array}$$

If $C \in \mathcal{C}^g(R)$ then the graded isomorphisms of degree zero :

$$(C \underset{R}{\otimes} R_i) \underset{R_i}{\otimes} R' \to C \underset{R}{\otimes} R', \qquad i = 1,2,$$

induce a natural isomorphism $\beta : F_1H_1 \to F_2H_2$. On the other hand one may construct the categorical fibre product

$$\mathcal{F}^g = \mathcal{C}^g(R_1) \underset{\mathcal{C}^g(R')}{\times} \mathcal{C}^g(R_2)$$

which consists of the triples (C_1,γ,C_2) with $C_i \in \mathcal{C}^g(R_i)$, $i = 1,2$, and with $\gamma : F_1C_1 \to F_2C_2$ an isomorphism in $\mathcal{C}^g(R')$. The morphisms in $\mathcal{F}^g$ are pairs $(q_1,q_2) : (C_1,\gamma,C_2) \mapsto (D_1,\gamma,D_2)$ where $q_i : C_i \to D_i$ is a $\mathcal{C}^g(R_i)$-morphism such that the following square is commutative :

$$\begin{array}{ccc} F_1C_1 & \xrightarrow{\gamma} & F_2C_2 \\ {\scriptstyle F_1q_1}\downarrow & & \downarrow{\scriptstyle F_2q_2} \\ F_1D_1 & \xrightarrow{\delta} & F_2D_2 \end{array}$$

We have canonical functors $G_i : \mathcal{F}^g \to \mathcal{C}^g(R_i)$, $i = 1,2$, defined by $(C_1,\gamma,C_2) \mapsto C_i$, $(q_1,q_2) \to q_i$, $i = 1,2$. Moreover, the following diagram is commutative modulo γ :

$$(**) \qquad \begin{array}{ccc} \mathcal{F}^g & \xrightarrow{G_1} & \mathcal{C}^g(R_1) \\ {\scriptstyle G_2}\downarrow & & \downarrow{\scriptstyle F_1} \\ \mathcal{C}^g(R_2) & \xrightarrow{F_2} & \mathcal{C}^g(R') \end{array} \qquad \gamma : F_1G_1 \to F_2G_2.$$

Here γ maps $F_1G_1(C_1,\gamma,C_2) = F_1C_1$ to $F_2G_2(C_1,\gamma,C_2) = F_2C_2$, explaining the notation. Since (**) is Cartesian square, there exists a unique functor

$T^g : C^g(R) \to F^g$ such that $H_i = G_i T^g$ and $\gamma T^g = \beta$, $i = 1,2$. Then $T^g C = (H_1 C, \beta, H_2 C)$ and $T^g f = (H_1 f, H_2 f)$. At this point we recall the definitions given in Chapter II, preceding Theorem II.19. From [60] recall then that a diagram like (**) is E-surjective if one of the functors F_1, F_2 is an E-surjective functor. Now we are ready to prove the following :

VI.4.5. THEOREM. *If in the Cartesian diagram of graded ring morphisms of degree zero (*) at least one of the morphisms* f_1 *or* f_2 *is a surjective map, then* $T^g : C^g(R) \to F^g$ *is an equivalence of categories.*

PROOF. (This comes down to the graded version of Milnor's theorem, cf. [60]). Looking at the proof of Milnor's theorem in loc. cit. one easily convinces oneself that the constructions there are indeed compatible with the gradation when it is available. So we only have to check whether T^g is cofinal and this will follow from the E-surjectivity of (*), i.e. we reduced the problem to the verification of the fact that either F_1 or F_2 is an E-surjective functor. More precisely, surjectivity of a graded ring morphism of degree zero, $f : R \to S$ say, entails E-surjectivity of $F = - \otimes_R S : C^g(R) \to C^g(S)$. A separate proof should now be given in each of the cases : $C^g = FP_g$, AZ_g, $\underline{Pic}_g$.

Let us provide a detailed proof for the situation $C^g = FP_g$, and we leave the proof for the case $C^g = \underline{Pic}_g$ to the reader. For convenience sake we write $A_R(-)$ for $Aut_{FP_g(R)}(-)$ and similar with respect to S. Now we have to establish that for any P in $FP_g(R)$ and any α in the commutator subgroup of $A_S(P \otimes_R S)$ there exist Q in $FP_g(R)$ and β in the commutator subgroup of $A_R(P \otimes_R Q)$ such that :

$$\beta \otimes_R S = \alpha \otimes_S FP = \alpha \otimes_S (S \otimes_R P) = \alpha \otimes_R P.$$

Clearly, it is natural to look for a Q of the form $P \oplus P'$ and since the gr-free R-modules of finite rank are cofinal in $FP_g(R)$ we may assume that P is of the form $\bigoplus_{i=1}^{r} R(n_i)$ and look for a Q of the form $P \oplus (\bigoplus_{j=1}^{s} R(m_j))$. Let $T = R(n_1,\dots,n_r)$ be the part of degree zero of the graded ring $M_r(R)(n_1,\dots n_r)$ with gradation defined as in the remarks after Proposition VI.3.3. Put $Gl_r(n_1,\dots,n_r{:}R) = Gl_r(T)$ and let

$$E_r(n_1,\ldots,n_r;\ R) \subset Gl_r(n_1,\ldots,n_r;\ R)$$

denote the elementary matrices i.e. the group (multiplicatively) generated by the set $\{I_r + \alpha e_{ij};\ 1 \leq i,\ j \leq r$ and $i \neq j$, $\alpha \in T$ is of the form ae_{kl} with $a \in R_{n_i - n_j}$, $1 \leq k$, $l \leq r$ and $k \neq l\}$, where I_r denotes the identity element of $M_r(R)$. We write $\underline{n}$ for $(n_1,\ldots,n_r)$.

If $b \in GL_r(\underline{n};R)$, say $b = I_r + q$, then we have :

$$\begin{pmatrix} b & 0 \\ 0 & b^{-1} \end{pmatrix} = \begin{pmatrix} I_r & q \\ 0 & I_r \end{pmatrix}\begin{pmatrix} I_r & 0 \\ I_r & I_r \end{pmatrix}\begin{pmatrix} I & -b^{-1}q \\ 0 & I \end{pmatrix}\begin{pmatrix} I & 0 \\ -b & I \end{pmatrix}$$

(because q and b commute).

If we write $q = \sum_{i,j} q_{ij} e_{ij}$, $1 \leq i,\ j \leq r$, with respect to the matrix idempotents e_{ij} then

$$\begin{pmatrix} I_r & q \\ 0 & I_r \end{pmatrix} = \prod_{i,j} \begin{pmatrix} I_r & q_{ij}e_{ij} \\ 0 & I_r \end{pmatrix}$$

and other terms in the expression for $\begin{pmatrix} b & 0 \\ 0 & b^{-1} \end{pmatrix}$ may be treated in a similar way so that finally it becomes clear that the latter matrix is an element of $E_{2r}(\underline{n};R)$. For arbitrary a,b in $Gl_r(\underline{n};R)$ we have that :

$$\begin{pmatrix} ab & 0 \\ 0 & I_r \end{pmatrix} = \begin{pmatrix} a & 0 \\ 0 & b \end{pmatrix}\begin{pmatrix} b & 0 \\ 0 & b^{-1} \end{pmatrix} \equiv \begin{pmatrix} a & 0 \\ 0 & b \end{pmatrix} \bmod E_{2r}(\underline{n};R)$$

and this is true for left as well as for right cosets module $E_{2r}(\underline{n};R)$. As a consequence of this, we obtain for arbitrary a,b in $Gl_r(\underline{n};R)$:

$$\left[\begin{pmatrix} a & 0 \\ 0 & I_r \end{pmatrix}, \begin{pmatrix} b & 0 \\ 0 & I_r \end{pmatrix}\right] = \begin{pmatrix} a^{-1}b^{-1} & 0 \\ 0 & I_r \end{pmatrix}\begin{pmatrix} ab & 0 \\ 0 & I_r \end{pmatrix}$$

$$= \varepsilon_1 \begin{pmatrix} a^{-1} & 0 \\ 0 & b^{-1} \end{pmatrix}\begin{pmatrix} a & 0 \\ 0 & b \end{pmatrix} \varepsilon_2 = \varepsilon_1\varepsilon_2 \in E_{2r}(\underline{n};R),$$

for certain $\varepsilon_1, \varepsilon_2 \in E_{2r}(\underline{n};R)$. The commutator subgroup of $Gl_r(\underline{n};R)$ is thus contained in $E_{2r}(\underline{n};R)$. Next, consider $A = I_r + e_{ij}\alpha \in E_r(\underline{n};R)$. Since $\alpha \in T$, we may write A as a commutator $[I_r + e_{ik}, I_r + \alpha e_{kj}]$ of elements in $E_r(\underline{n};R)$. Therefore, $E_r(n;R)$ is equal to its commutator subgroup. Embed $Gl_m(\underline{n};R)$ into $Gl_{m+s}(\underline{n};R)$ by mapping α to $\alpha \oplus I_s$ and similarly : $E_m(\underline{n};R) \to E_{m+s}(\underline{n};R)$. In this way we obtain a directed system and it makes sense to define

$$Gl(n;R) = \varinjlim_m Gl_m(\underline{n};R), \qquad E(\underline{n};R) = \varinjlim_m Gl_m(\underline{n};R).$$

By the preceding argumèntation it is clear that $E(\underline{n};R)$ is its own commutator subgroup and also the commutator subgroup of $Gl(\underline{n};R)$. In the case we are considering i.e. $C^g = FP_g$, $P = \bigoplus_{i=1}^{r} R(n_i)$, we have that $A_S(P \otimes_R S)$ is just the group $Gl_1(\underline{n};S)$. For any $m \in \mathbb{N}$, $A_R(P^m) \cong Gl_m(\underline{n};R)$. So we finally have reduced the problem to the following : for an α in the commutator subgroup of $Gl_1(\underline{n};S)$ find an $n \in \mathbb{N}$ and a β in the commutator subgroup of $Gl_n(\underline{n};R)$ such that $\beta \equiv \alpha \oplus I$ modulo Ker(f). Surjectivity of f yields that :

$$\begin{array}{ccc}
[Gl(\underline{n};R),\ Gl(\underline{n};R)] & \to & [Gl(\underline{n};S),\ Gl(\underline{n};S)] \\
\downarrow = & & \downarrow = \\
E(\underline{n};R) & \longrightarrow & E(\underline{n};R)
\end{array}$$

is a surjective morphism. Obviously this is then true at each level i.e. the induced maps $E_m(\underline{n};S) \to E_m(\underline{n};S)$ are indeed surjective and this proves our claims. □

VI.4.6. THEOREM. *Consider a diagram of graded ring morphisms of degree zero, where* f_1 *or* f_2 *is supposed to be surjective :*

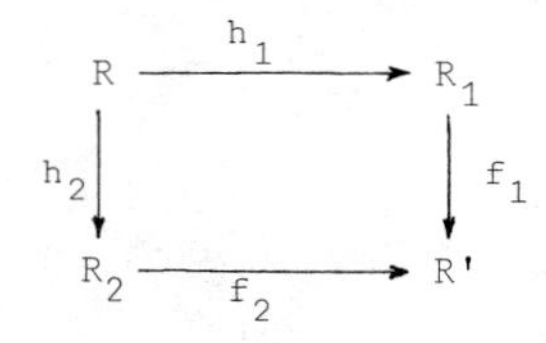

If R' *is a gr-semilocal ring, then we have the following exact Mayer-Vietoris sequence :*

$$0 \to Br^g(R) \to Br^g(R_1) \oplus Br^g(R_2) \to Br^g(R').$$

PROOF. The diagram given induces a sequence of group homomorphisms :

$$Br^g(R) \to Br^g(R_1) \oplus Br^g(R_2) \to Br^g(R'),$$

where the first map is the diagronal morphism. By Milnor's theorem we have a Cartesian diagram :

$$\begin{array}{ccc} AZ_g(R) & \xrightarrow{H_1} & AZ_g(R_1) \\ {\scriptstyle H_2}\downarrow & & \downarrow \\ AZ_g(R_2) & \xrightarrow[F_2]{} & AZ_g(R') \end{array}$$

$$\beta : F_1H_1 \to F_2H_2,$$

where $F_i = - \underset{R}{\otimes} R_i$, $H_i = - \underset{R}{\otimes} R'$.
The sequel of the proof is now a modification of a similar result in [24]. A graded Azumaya algebra A may be viewed as (A_1,α,A_2) in the fibre product. Assume that $[A] \in Br^g(R)$ is in the kernel of the morphism $Br^g(R) \to Br^g(R_1) \oplus Br^g(R_2)$, i.e. $[G_iA] = [A_i] = 0$ in $Br^g(R_i)$, $i = 1,2$. Hence there exist $P_i \in FP_g(R_i)$ such that $A_i = END_{R_i}(P_i)$, $i = 1,2$. The isomorphism of degree zero :

$$END_{R'}(F_1P_1) = F_1\ END_{R_1}(P_1) = F_1A_1 \to F_2A_2 = F_2\ END_{R_2}(P_2) = END_{R'}(F_2P_2)$$

is induced by an R'-isomorphism of degree zero :

$$f : F_1P_1 \underset{R'}{\otimes} I \to F_2P_2,$$

for some $I \in \underline{Pic}_g(R')$. Actually, $I = HOM_S(F_1P_1,F_2P_2)$ where $S = END_{R'}(F_1P_1)$, as one may easily verify. Since R' is gr-semilocal, $I = R'(n)$, and thus $f : (F_1P_1)(n) = F_1(P_1(n)) \to F_2P_2$. But

$$END_R(P(n)) = END_R(P)$$

is a general fact, for any graded ring R and graded R-module P, so it follows that $A = (A_1,\alpha,A_2) = END_R(P)$ where $P = (P_1(n),f,P_2)$ is a graded projective R-module which is finitely generated. This establishes exactness of the sequence $0 \to Br^g(R) \to Br^g(R_1) \oplus Br^g(R_2)$. Select $([A_1],[A_2])$ in $Br^g(R_1) \oplus Br^g(R_2)$ in the kernel of the difference morphism. Obviously, it will not be restrictive to suppose that A_1 and A_2 have constant rank. From $[F_1A_1] = [F_2A_2]$ in $Br^g(R')$ it follows that

$$F_1A_1 \underset{R'}{\otimes} END_{R'}(P') = F_2A_2 \underset{R'}{\otimes} END_{R'}(Q')$$

for certain $P',Q' \in FP_g(R')$. Again it is not restrictive to assume that both P' and Q' have constant graded rank; P' and Q' are gr-free R-modules, say $P' = \bigoplus_{i=1}^{r} R'(n_i)$, $Q' = \bigoplus_{j=1}^{s} R(m_j)$. Define

$$P_1 = \bigoplus_{i=1}^{r} R_1(n_i), \quad Q_2 = \bigoplus_{j=1}^{s} R_r(m_j).$$

Then $F_1P_1 = P'$, $F_2Q_2 = Q'$ and also :

$$F_1(A_1 \underset{R_1}{\otimes} END_{R_1}(P_1)) = F_1A_1 \underset{R'}{\otimes} END_{R'}(P') =$$

$$= F_2A_2 \underset{R'}{\otimes} END_{R'}(Q') = F_2(A_2 \underset{R_2}{\otimes} END_{R_2}(Q_2)).$$

The latter implies that $([A_1], [A_2])$ is in the image of the diagonal morphism. Conversely, if $([A_1], [A_2])$ is in the image of the diagronal morphism then it is in the kernel of the difference morphism, as one easily checks, so the theorem is proved. □

VI.4.7. <u>THEOREM</u>. (In the situation of foregoing theorem). *There is a long exact sequence :*

$$0 \to U(R) \to U(R_1) \oplus U(R_2) \to U(R') \to$$
$$\to Pic_g(R) \to Pic_g(R_1) \oplus Pic_g(R_2) \to Pic_g(R') \to$$
$$\to Br^g(R) \to Br^g(R_1) \oplus Br^g(R_2) \to Br^g(R').$$

<u>PROOF</u>. Combination of Mayer-Vietoris sequences and some technicalities. Note that the hypothesis "R' is gr-semilocal" may be weakened here ! We do not go into the details. □

VI.4.8. <u>Note</u>. Over a gr-Dedekind ring, one may check that isomorphic graded Azumaya algebras are also equivalent in the graded Brauer group.

This follows from the fact that this same statement is true for a gr-field and the method of proof of Proposition VI.3.11.

We conclude this section by giving an important example strongly related to the geometrical theory of Chapter VII.

If S is a commutative $\mathbb{Z}$-graded domain then $\overline{S}$ will stand for the integral closure of S in its field of fractions Q(S). The conductor c of $\overline{S}$ in S is a common graded ideal of S and $\overline{S}$. Now consider a gr-local Noetherian domain R and let c be the conductor of $\overline{R}$ in R, then there is an exact sequence :

$$0 \to Br^g(R) \to Br^g(\overline{R}) \oplus Br^g(R/c) \to Br^g(\overline{R}/c).$$

To prove this it suffices to note that $\overline{R}/c$ is gr-semilocal and then one applies Theorem VI.4.6. to the Cartesian square :

$$\begin{array}{ccc} R & \to & \overline{R} \\ \downarrow & & \downarrow \\ R/c & \to & \overline{R}/c \end{array}$$

VI.4.9. <u>LEMMA</u>. *Let* S *be any commutative* $\mathbb{Z}$*-graded ring and let* I *be a nilpotent ideal of* S *which is graded, then the canonical morphism* $Br^g(S) \to Br^g(S/I)$ *is an isomorphism.*

<u>PROOF</u>. A faithful reconstruction of the proof for the ungraded equivalent statement presented in [31]. Along the way one uses : the graded version of Nakayama's lemma, the graded version of a theorem of H. Bass, cf.[12], and the graded version of Theorem 1 in [31]. □

In the sequel of this section R is a positively graded Noetherian domain such that dim R ⩽ 2 and which is generated as an R_0-algebra by the elements of degree one; furthermore we assume that R_0 is a field and we write $k = R_0$. As before $\overline{R}$ is the integral closure of R and c is the conductor of $\overline{R}$ in R. Clearly $c \subset R_+$ and rad(c) is a graded ideal such that $(\mathrm{rad}(c))^n \subset c$ for some suitable $n \in \mathbb{N}$. The assumption on the dimension of R entails that the closed set $V_+(c)$ in Proj(R) contains only a finite number of points, say $V_+(c) = \{P_1,\ldots,P_m\}$. We consider the inclusion $R/\mathrm{rad}(c) \to \bigoplus_{i=1}^m R/P_i$ and the conductor d of $\bigoplus_{i=1}^m R/P_i$ in $R/\mathrm{rad}(c)$. Actually $d = \bigoplus_{i=1}^m d_i$ where $d_i = (\bigcap_{j\neq i} P_j) \bmod P_i$. So we obtain a Cartesian square :

$$\begin{array}{ccc} R/\mathrm{rad}(c) & \longrightarrow & \bigoplus_{i=1}^m R/P_i = S \\ \downarrow & & \downarrow \\ (R/\mathrm{rad}(c))/d & \longrightarrow & (\bigoplus_{i=1}^m R/P_i)/\bigoplus_{i=1}^m d_i \end{array}$$

Since $(\bigoplus_{i=1}^m R/P_i)/\bigoplus_{i=1}^m d_i$ is gr-semilocal, it is a direct consequence of Theorem VI.4.6., that we have the following Mayer-Vietoris sequence :

$$0 \to \mathrm{Br}^g(R/\mathrm{rad}(c)) \to \bigoplus_{i=1}^m \mathrm{Br}^g(R/P_i) \oplus \mathrm{Br}^g((R/\mathrm{rad}(c))/d) \to$$

$$\xrightarrow{\eta} \mathrm{Br}^g((\bigoplus_{i=1}^m R/P_i)/\bigoplus_{i=1}^m d_i).$$

Furthermore, since $\mathrm{rad}(c)/c$ is nilpotent, Lemma VI4.9. entails that $\mathrm{Br}^g(R/c) = \mathrm{Br}^g(R/\mathrm{rad}(c))$ i.e. $\mathrm{Br}^g(R/c) = \mathrm{Ker}\ \eta$. Another consequence of the fact that $\dim R \leqslant 2$ is that $\bigcap_{j\neq i} P_j \not\subset P_i$ hence $\mathrm{rad}(P_i+d_i) = R_+$. So we obtain :

$$\mathrm{Br}^g(S/d) = \bigoplus_{i=1}^m \mathrm{Br}^g(R/P_i + \bigcap_{j\neq i} P_j) = \bigoplus_{i=1}^m \mathrm{Br}^g(R/\mathrm{rad}(P_i + \bigcap_{j\neq i} P_j))$$

$$= \bigoplus_{i=1}^m \mathrm{Br}^g(R/R_+) = (\mathrm{Br}(k))^m.$$

In a similar way we obtain :

$$\mathrm{Br}^g((R/\mathrm{rad}(c))/d) = \mathrm{Br}^g(R/\Sigma_i(\bigcap_{j\neq i} P_j))$$

$$= \mathrm{Br}^g(R/\mathrm{rad}(\Sigma_i(\bigcap_{j\neq i} P_j))) = \mathrm{Br}^g(R/R_+) = \mathrm{Br}(k).$$

The Mayer-Vietoris sequence mentioned before now reduces to :

$$0 \to Br^g(R/c) \xrightarrow{i} \bigoplus_{i=1}^{m} Br^g(R/P_i) \oplus Br(k) \xrightarrow{\eta} (Br(k))^m$$

where the morphisms i and η are defined by : $i(\alpha) = (\alpha_1,\ldots,\alpha_m,\alpha')$ where α_j, $j = 1,\ldots,m$, is the image of α under the canonical morphism $\pi_j : Br^g(R/c) \to Br^g(R/P_j)$, and where α' is the image of α under $Br^g(R/C) \to Br^g(R/R_+) = Br(k)$; $\eta(\beta_1,\ldots,\beta_m,\beta') = (\beta_1'-\beta',\ldots,\beta_m'-\beta')$, where β_j, $j = 1,\ldots,m$, is the image of β_j under the morphism

$$Br^g(R/P_j) \to Br^g(R/R_+) = Br(k).$$

Consequently, $(\alpha_1,\ldots,\alpha_m,\alpha')$ is in the image of $Br^g(R/c)$ if and only if $\alpha_j' = \alpha'$, $j = 1,\ldots,m$. Define $B^g(R,c)$ to be the subgroup of $\bigoplus_{P\in V_+(c)} Br^g(R/P)$ consisting of the equivalence classes given by an element $(x_P,\ P \in V_+(c))$ where $\eta_P(x_P) = \eta_{P'}(x_{P'})$ for all $P,P' \in V_+(c)$, $\eta : Br^g(R/P) \to Br^g(R/R_+)$ being the canonical morphism. Similarly we may define $B^g(\overline{R},c)$ as the subgroup of $\bigoplus_{Q\in V_+(c)} Br^g(\overline{R}/Q)$ (where $\overline{V}_+(c)$ is the closed set defined by c in $Proj(\overline{R})$) consisting of the $(y_Q,\ Q \in \overline{V}_+(c))$ such that $\xi_Q(y_Q) = \xi_{Q'}(y_{Q'})$ for all $Q,Q' \in \overline{V}_+(c)$, where ξ_Q is the canonical morphism, $Br^g(\overline{R}/Q) \to Br^g(\overline{R}/\overline{R}_+)$. We can sum up all this in :

VI.4.10. THEOREM. *Let* R *be a positively graded Noetherian domain with* $\dim R \leq 2$ *such that* R *is generated as an* R_0*-algebra by* R_1 *and such that* $R_0 = k$ *is a field, then the graded Brauer group* $Br^g(R)$ *is completely described by the following exact sequence :*

$$0 \to Br^g(R) \to Br^g(\overline{R}) \oplus B^g(R,c) \to B^g(\overline{R},c).$$

VI.4.11. Remarks. In the foregoing we implicitly used the fact that $\overline{R}$ is again positively graded. This may be checked in a straightforward manner, actually we come back to this in Chapter VII. On the other hand it is not obvious that $\overline{R}/\overline{R}_+ = k$, and indeed it need not be true at all. As an example one may consider $R = \mathbb{R}[X,Y,Z]/(X^2-XY+Y^2)$. It is clear that $t = XY^{-1} \bmod (X^2-XY+Y^2)$ in the functionfield of R satisfies the monic equation $t^2 -t+1 = 0$, yet $t \notin k =$ and deg $t = 0$. In this case $\overline{R}/\overline{R}_+ = \mathbb{C}$.

VI.5. Graded Brauer Groups of Certain Graded Rings.

In this section we study the graded Brauer group of a gr-local ring. As in the ungraded case the main tool in describing, up to Brauer gr-equivalences, the algebras representing elements of the graded Brauer group of a gr-local ring will be Galois cohomology and crossed product theorems. The construction of a Galois gr-splitting ring for a given Azumaya algebra over a gr-local ring cannot be obtained by straightforward "gradation" of the ungraded analogues; this is partly due to some of the peculiarities of graded nature which are inherent to the definition of gr-splitting. The results in this section could be obtained in a somewhat shorter way if one developes first the graded (Amitsur-)cohomological theory expounded now in Section VI.6. We have chosen to build up the theory step by step starting from the special case of the graded Brauer group of a gr-discrete valuation ring, adapting the methods used to the more general case of Noetherian gr-local rings and finally, introducing graded Henselisation in order to replace the technique of graded completion, we deal with general gr-local rings. In our opinion this way of presenting the material is the most natural; there is however a danger of introducing consecutive generalizations of the same basic idea in the method of proof. We hope to have avoided this flaw by focussing on the theoretical novelties appearing at each consecutive step of the generalization process, while some of the straightforward routine verifications, along the lines of proofs already encountered when dealing with some special case studied before, are left to the reader.

In the first part of this section we consider a gr-Dedekind ring D; see Section VI.1. and also Section VI.3. for definitions and properties of gr-Dedekind rings. Let us just recall here that D is then a regular domain of global dimension at most two. If P is a graded prime ideal of D then $Q^g_{D-P}(D)$ is a gr-discrete valuation ring of K^g, the gr-field of fractions of D. From Theorem I.3.10. it follows that $Br(D) = \cap\{Br(Q_{D-P}(D));\ P \in X^{(1)}(D)\}$. As a direct consequence of Corollary VI.3.12., or immediately from Theorem VI.3.25. we also have :

$$Br^g(D) = \cap\{Br^g(Q^g_{D-P}(D));\ P \in X^{(1)}_g(D)\}.$$

If for $P \in X^{(1)}(D)$ we write e_P for the ramification index of the (graded) valuation associated to $Q^g_{D-P}(D)$ with respect to the valuation associated to $(Q^g_{D-P}(D))_0$, then e_P is exactly the maximal positive integer n such that $D_1 \subset P^n$, and it is also the minimal positive integer m such that $(Q^g_{D-P}(D))_m$ contains a unit.

VI.5.1. LEMMA. *Any graded Azumaya algebra* A *over a discrete gr-valuation ring* D *is gr-equivalent to a graded Azumaya algebra over* D *which is a domain.*

PROOF. The graded ring of fractions $Q^g(A) = A \underset{D}{\otimes} K^g$ is a gr-c.s.a. hence we may assume it is of the form $M_n(\Delta[X,X^{-1},\varphi])(\underline{d})$ (cf. Proposition VI.3.3. and consequent remarks) for some $\underline{d} \in \mathbb{Z}^n$. Let B be any graded maximal D-order in the gr-field $\Delta[X,X^{-1},\varphi]$, and identify $M_n(B)(\underline{d}) = \Lambda$ with a maximal graded order in $Q^g(A)$. Let $I_c = \{q \in Q^g(A),\ q\Lambda \subset A\}$ be the conductor of Λ in A. Clearly, I_c is a graded left ideal of A and a graded right ideal of Λ. Write ω for the unique gr-maximal ideal of D and let $W = A\omega$ be the unique gr-maximal ideal of A lying over ω. For any graded left ideal L of A we have that L/WL is a graded left ideal of $A/W \cong M_r(\Delta_1[Y,Y^{-1},\psi])(\underline{d})$ where Δ_1 is a skewfield, ψ an automorphism of Δ_1. So L/WL is generated as an A/W-module by a single homogeneous element, $\overline{y}$ say. Pick a representative $y \in L$ for $\overline{y}$. Then $L = Ay + WL$ (note that both A and L are gr-free D-modules while L is a direct summand of A, therefore $W \cap L = WL$) and the graded version of Nakayama's lemma entails $L = Ay$. Applying this with $L = I_c$, it follows that $I_c = Au$ for some homogeneous u. Since we have $A \underset{D}{\otimes} K^g$ it follows that $I_c \cap D \neq 0$ and thus there exists a homogeneous $a \in h(A)$ such that au is a unit in $Q^g(A)$; therefore u is a unit in $Q^g(A)$. So we obtain that $\Lambda \subset u^{-1}Au$. But $u^{-1}Au$ is a graded D-order in $Q^g(A)$, so the maximality hypotheses entail that $\Lambda \cong u^{-1}Au$. Finally, since $M_n(B)(\underline{d})$ is a graded Azumaya algebra, B is a graded Azumaya algebra too and A is obviously gr-equivalent to B. That B is a domain is evident from $B \subset \Delta[X,X^{-1},\varphi]$. □

The following proposition is essential in solving the crossed product problem over discrete gr-valuation rings. The proof may be considered to be a graded version of a proof given by M. Auslander, O. Goldman in [10], which they credit J-P. Serre for. The problems of graded nature which force some modifications in the original ungraded proof are mainly due to the fact that a discrete valuation ring D need not necessarily be of the form $D_0[T,T^{-1}]$ with deg T = 1 !

VI.5.2. PROPOSITION. *Let D be a discrete gr-valuation ring. Every class* $\alpha \in Br^g(D)$ *may be represented by a graded Azumaya algebra over D, containing a maximal commutative subring S with the following properties :*

1. *S is a graded domain and a gr-free D-module of rank* $\sqrt{[A:D]}$.
2. *S is of the form* $S_0 \underset{D_0}{\otimes} D$ *where* S_0 *is a separable extension of* D_0. *Consequently S is gr-semilocal and actually S is a gr-principal ideal domain.*

PROOF. Let B be a graded Azumaya algebra representing $\alpha \in Br^g(D)$ and write $W = B\omega$, $\overline{B} = B/W$ and $m^2 = [B : D]$. The residue gr-field of D is denoted by $\overline{D} = k[T,T^{-1}]$. Since $\overline{B}$ is a graded Azumaya algebra of rank m^2 over $k[T,T^{-1}]$ we may apply Proposition VI.3.7. to deduce that $\overline{B}$ may be gr-split by $l[T,T^{-1}]$ for some separable finite field extension l/k. Write $l = k(\theta)$, deg θ = 0, and let $f \in k[X]$ be the minimal polynomial satisfied by θ over k. Consider a monic polynomial $F \in D_0[X]$ having the same degree in X as f and such that F mod $\omega[X]$ = f; F may be obtained by lifting the coefficients of f to D_0. Since D_0 is a U.F.D. and F is irreducible it follows that (F) is a prime ideal of $D_0[X]$. Hence $T_0 = D_0[X]/(F)$ is a separable D_0-algebra which is a domain and also a free D_0-module of rank $\deg_X F = \deg_X f$. It is a direct consequence of the foregoing facts that T = D[X]/(F) is a graded separable D-algebra which is also a gr-free D-module of rank $\deg_X f$. In order to show that T is actually a domain it will suffice to establish that T has no homogeneous zero divisors. Define the gradation on D[X] by giving X degree zero and suppose $u,g,h \in h(D[X])$ are such that uF = gh. As pointed out earlier, D contains an invertible element of degree e, say u_e. Pick nonzero elements $d_1 \in D_{e-\deg g}$, $d_2 \in D_{e-\deg h}$. (Note that we may assume $D_n \neq 0$ for all $n \in \mathbb{Z}$!). Then : $(u_e^{-1} d_1 g)(u_e^{-1} d_2 h) = (u_e^{-2} d_1 d_2 u)F$. Since F generates a prime ideal of $D_0[X]$, either $u_e^{-1} d_1 g$ or $u_e^{-1} d_2 h$ is in $D_0[X]F$, say $u^{-1} d_1 g = v_0 F$, $v_0 \in D_0[X]$. If $d_1 \notin \omega$ then $g \in D[X]F$ while on the other hand, $d_1 \in \omega$ yields $u_e^{-1} d_1 \in \omega$ and thus $v_0 F \in \omega[X]$. Now $\overline{D}[X] = k[X,T,T^{-1}]$ is prime and $F \notin \omega[X]$, hence $v_0 \in (\omega[X])_0$. Let π denote the homogeneous generator for ω, then $v_0 = \pi^{\nu} v_0^1$, $u_e^{-1} d_1 = \pi^{\nu}\lambda$ with $v_0^1 \in D[X] - \omega[X]$ and $\lambda \in D - \omega$. Therefore $\lambda g = v_0^1 F$ but then $\lambda^{-1} \in D$ yields $g \in D[X]F$. Next consider the commutative diagram :

$$(*) \qquad \begin{array}{ccccccc} D & \hookrightarrow & T & \hookrightarrow & D^{cl} & \hookrightarrow & Q(T) \\ \downarrow & & \downarrow & & & & \\ \overline{D} & \hookrightarrow & l[T,T^{-1}] & & & & \end{array}$$

where D^{cl} is the integral closure of D in the field of fractions Q(T) of T. It is obvious that D^{cl} is also the integral closure of T and as the latter is a graded ring it follows that D^{cl} is a graded ring too. The graded version of the Krull-Akizuki theorem, (Proposition VI.1.9.) states that D^{cl} is a gr-Dedekind ring. We have established :

$$[Q(T) : K] = [T : D] = [l : k]$$

and therefore $D^{cl}/D^{cl}\omega = l[T,T^{-1}]$ while $D^{cl}\omega \subset J^g(D^{cl})$. Separability of Q(T) over Q(D) = K entails that D^{cl} is a finitely generated D-module and so we may apply the graded version of Nakayama's lemma and obtain $D^{cl} = T$. By construction T is gr-local, so from $T = D^{cl}$ it then follows that T is a discrete gr-valuation ring. We have reached the situation where $B \otimes_D T$ is a graded Azumaya algebra over the discrete gr-valuation ring T, such that $B \otimes_D T/(B \otimes_D T)\omega = \overline{B} \otimes_{\overline{D}} \overline{T}$ is a graded matrix ring over $\overline{T}$. So we may assume now that B is such that $\overline{B} = M_m(k[T,T^{-1}])(\underline{d})$, $d \in \mathbb{Z}^m$. By Lemma VI.5.1. there exists a graded Azumaya algebra A over D which is gr-equivalent to B and such that A has no zero divisors. Actually from the proof of Lemma VI.5.1. we retain that $B = M_s(A)(\underline{e})$, $\underline{e} \in \mathbb{Z}^s$ for some $s \in \mathbb{N}$. Then $\overline{B} = M_s(\overline{A})(\underline{e}) = M_m(k[T,T^{-1}])(\underline{d})$ entails : $\overline{A} \underset{gr}{\sim} 1$, i.e. $\overline{A} = M_t(k[T,T^{-1}])(\underline{f})$ for some $t \in \mathbb{N}$, $\underline{f} \in \mathbb{Z}^t$. In $M_t(k)$ we find a commutative subring $k[\alpha]$ such that $[k[\alpha] : k] = t$, e.g. let α be a diagonal matrix with different entries (nonzero). Consider $k[T,T^{-1}][\alpha]$ and lift α to $\beta \in A_0$. It is obvious that $D[\beta]$ is a commutative subring of A and also a domain. Moreover, $K[\beta]$ is a commutative subring in the skewfield $Q(A)$. Hence, from $[K[\beta] : K] \leqslant t = \sqrt{[A:D]}$ and the fact that $\{1,\beta,\ldots,\beta^{t-1}\}$ is D-independent, it follows that $\beta^t = \sum_{i=0}^{t-1} k_i\beta^i$ with $k_i \in K$. Multiplying this equation by suitable common denominators and using the graded structure of B we obtain from this that $d\beta^t = \sum_{i=0}^{t-1} d_i\beta^i$ with $d,d_i \in D$ and homogeneous. Moreover, $\deg \beta = 0$ entails $\deg d = \deg d_i$ for all i. If $d \in \omega$ then $0 = \sum_{i=0}^{t-1} \bar{d}_i\bar{\beta}^i = \sum_{i=0}^{t-1} \bar{d}_i\alpha^i$ implies $d_i \in \omega$ for all i. Consequently, the assumption $d \in \omega$ leads to simplification by π and after a finite number of similar steps we obtain a relation : $d'\beta^t = \sum_{i=0}^{t-1} d_i'\beta^i$ with $d',d_i' \in h(D)$ and $d' \notin \omega$. But then it is clear that d' is a unit of D and therefore $\beta^t \in D + D_\beta + \ldots + D\beta^{t-1}$. This states exactly that the free D-module S generated by $\{1,\beta,\ldots,\beta^{t-1}\}$ is a subring of A and consequently a domain as well. Put

$$H = X^t - \sum_{i=0}^{t-1} (d')^{-1} d_i' X^i.$$

Obviously H is a lifting of the minimal polynomial of α over k and it is clear that $S = D[X]/D[X]H = D_0[X]/D_0[X]H \underset{D_0}{\otimes} D$. The properties of S listed in the statement of the proposition are easily verified. That S gr-splits A is a direct consequence of the fact that the field of fractions $Q(S)$ of S splits $Q(A)$ (by a dimension argument) combined with the fact that $Br^g(S) \to Br(Q(S))$ is an injection. Moreover since S is a gr-semilocal gr-Dedekind ring it is also a gr-principal ideal ring. □

The above proposition may be strengthened to :

VI.5.3. <u>THEOREM</u>. *Let* D *be a discrete gr-valuation ring. Every* $\alpha \in Br^g(D)$ *may be represented by a graded Azumaya algebra* A *over* D *containing a maximal commutative subring* S *with the following properties :*

1. S *is a graded Galois extension of* D.
2. S *is a gr-principal ideal domain, hence* $Pic(S) = Pic^g(S) = 1$.
3. $S = S_0 \underset{D_0}{\otimes} D$.

4. *The units of* S *are homogeneous elements.*

PROOF. Let A_1 and S_1 be as A and S obtained in Proposition VI.5.2. Since S_1 does not contain idempotent elements different from 0 and 1 it follows, cf. [68], that the normalization S of S_1 does not contain idempotent elements different from 0 and 1. Let G be the Galois group of S over D. In Remark III.2.3. it has been pointed out that the trace, $tr : S \to C$, given by $tr(x) = \sum_{\sigma \in G} \sigma(x)$, is a gree generator of the right S-module $Hom_D(S,D) = HOM_D(S,D)$. If $x \in S$ is nilpotent then tr(xs) is nilpotent in D for all $s \in S$, i.e. $tr.x = 0$ in $Hom_D(S,D)$. Since tr is a free generator, $x = 0$ follows. This shows that S is reduced (semiprime). Therefore we may apply Lemma VI.2.5. and deduce that $U(S) \subset h(S)$. Since $S_1 = (S_1)_0 \underset{D_0}{\otimes} D$ it is clear that S may be obtained by constructing a D_0-normalization of $(S_1)_0$ and tensor it by D over D_0. In view of the above remarks we obtain that S_0 has no nontrivial idempotents, the ring of fractions L of S_0 is a Galois ring extension of K, hence L is semisimple say $L = L_1 \oplus \ldots \oplus L_r$. We have a commutative diagram :

$$(*) \qquad \begin{array}{ccccc} D_0 & \hookrightarrow & S_0 & \hookrightarrow & D_0^{cl} = (D_0^{cl})_1 \oplus \ldots \oplus (D_0^{cl})_r \\ \downarrow & & \downarrow & & \\ k & \hookrightarrow & \overline{S_0} & = & S_0/S_0\omega_0 \end{array}$$

where D_0^{cl} is the integral closure of D_0 in L (this of course splits as the direct sum of the integral closures of D_0 in the field $L_1,\ldots,L_r$), and where $\overline{S}_0$ has Galois group G over k. From the dimension equality $[L : K] = |G| = [\overline{S}_0 : k]$ we deduce that $D_0^{cl} = S_0 + D_0^{cl}\omega_0$ and from Nakayama's lemma $S_0 = D_0^{cl}$ follows. The only idempotents of S_0 are 0 and 1, so $r = 1$ and S_0 is a domain. The Krull-Akizuki theorem learns that D_0^{cl} is a Dedekind ring and it is a finitely generated module over the discrete valuation ring D_0 i.e. a principal ideal domain. One easily verifies that S cannot have homogeneous zero divisors, hence S has no zero divisors at all (using the fact that S is reduced !) Knowing that S is a domain we are now in a position to continue the proof in exactly the same way as in the proof of the foregoing proposition, we leave this verification to the reader. □

VI.5.4. COROLLARY. *Every graded Azumaya algebra over a discrete gr-valuation ring is gr-equivalent to a crossed product* $S[u_\sigma;\ \sigma \in G]$ *where* S *is a gr-principal ideal domain and a Galois extension of* D. □

VI.5.5. Remarks.

1. In the proof of Theorem VI.5.3. we implicitly used the fact that a normal closure of a graded ring S_1 may be made into a graded ring containing S_1 as a graded subring. That this is indeed true may be verified by checking along the lines of proof of Theorem 2.9. in [68] (or the implication 1 ⇒ 2 on p. 89 of [68]).

2. The particularities of the arithmetically graded case considered above may be summed up in the following :

a) upto gr-equivalence we may assume that graded Azumaya algebras have no zerodivisors,

b) the construction cannot be carried out completely in degree zero but a substantial part of it is strongly related to constructions in degree zero.

We now turn to more general gr-local rings i.e. Noetherian (strictly-) gr-local rings. In this step we loose property a) mentioned above and due to this fact we have to introduce graded completions. On the other hand the link to "degree zero constructions" will prevail. Let us first introduce the technical tool of graded completion (abbreviated : gr-completion).

Consider a graded ideal I of R. The I-adic valuation on R may be defined by $v(0) = +\infty$ and $v(x) = \max\{n \in \mathbb{N};\ x \in I^n\}$ for $x \neq 0$ in R. Put $d(x,y) = \alpha^{v(x-y)}$ where α is any fixed element of the open interval $]0,1[$ in $\mathbb{R}$, for $x,y \in R$. With this pseudo-ultrametric d, R becomes a pseudo-ultrametric space which is a Hausdorff space if and only if $\bigcap_n I^n = 0$. The zero-dimensional topology generated by d is just the I-adic topology of R. We let v^0 be the I_0-adic valuation on R_0.
The ring R is said to be I-complete if it is complete in the topological

sense for the I-adic topology or the associated ultrametric d. If R is not complete then one constructs the <u>completion</u> $\hat{R}$ e.g. by the classical Cauchy sequence method, cf. [15], [116]. Unfortunately if I is a graded ideal of a graded ring R then the I-adic completion $\hat{R}$ need not be a graded ring e.g. $R = k[X]$ where k is a field, $I = (X)$, then $\hat{R} = k[[X]]$ is not graded. So we proceed in a somewhat different way in order to define the gr-completion. For $r \in \mathbb{Z}$, consider the restriction v_r of v to R_r, i.e. $v_r(x) = \max\{n \in \mathbb{N};\ x \in (I^n)_r\}$ for $x \in R_r$ and define v^g for $x \in R$ by $v^g(x) = \min\{v_r(x_r);\ r \in \mathbb{Z}\}$ where the x_r are the homogeneous components of x of degree r. Note that in general $v^0 \neq v_0$.
For each map $f : \mathbb{Z} \to]0,1[$ define v_f on R by $v_f(x) = \min\{v_r(x_r)f(r);\ r \in \mathbb{Z}\}$.
For any $\alpha \in]0,1[$, $d_f(x,y) = \alpha^{v_f(x-y)}$ determines a new ultrametric on R and the family $\{d_f;\ f \in]0,1[^{\mathbb{Z}}\}$ generates a uniformity $\mathcal{U}_I$-gr. Details about uniformities may be found in [116]; one easily verifies that $\mathcal{U}_I$-gr is non-Archimedean.
We say that R is <u>gr-I-complete</u> if R is complete for the uniformity $\mathcal{U}_I$-gr. In general, the gr-I-adic completion $\hat{R}^g$ may be constructed by the Cauchy net method. Let $\hat{R}_r$ be the completion of R_r with respect to v_r, $r \in \mathbb{Z}$. Clearly, $\hat{R}^g$, $\hat{R}$, $\hat{R}_0$ are rings and each $\hat{R}_r$ is an $\hat{R}_0$-module. It is straightforward to verify that

$$\hat{R} = \{(x_r,\ r \in \mathbb{Z}) \in \prod_{r \in \mathbb{Z}} \hat{R}_r;\ \lim_{|r| \to \infty} v_r(x_r) = \infty\}.$$

VI.5.6. <u>LEMMA</u>. *A sequence* $\{x^{(n)}, n\} \subset R$ *is a* $\mathcal{U}_I$*-gr-Cauchy sequence if and only if the following properties hold :*

1. $\{x_r^{(n)}, n\} \subset R_r$ *is a* v_r*-Cauchy sequence for all* $r \in \mathbb{Z}$.
2. *There is a finite subset* F *of* $\mathbb{Z}$ *such that for all* $n \in \mathbb{N}$, $x^{(n)} \in \bigoplus_{r \in F} R_r$.

<u>PROOF</u>. Cf. [17]. □

Another result from loc. cit. which we evoke here is :

VI.5.7. <u>PROPOSITION</u>. *Let* I *be a graded ideal of the* $\mathbb{Z}$*-graded ring* R*, then* $\hat{R}^g = \bigoplus_{r \in \mathbb{Z}} \hat{R}_r$. □

From these results one now easily derives the following description of the gr-I-adic completion, which makes the foregoing analytically flavoured results superfluous in the sense that one may use next proposition as the definition of the gr-I-adic completion.

VI.5.8. PROPOSITION. *Let* I *be a graded ideal of a graded ring* R. *The* I-*adic completion* $\hat{R}$ *is a filtered ring and the gr*-I-*adic completion* $\hat{R}^g$ *is nothing but the associated graded ring of the filtration on* $\hat{R}$. *Furthermore* $\hat{R}^g$ *is a graded faithfully flat* R-*module. If* I *is a gr-maximal ideal then* $\hat{R}^g$ *is a gr-local ring.*

PROOF. The first statements follow easily from the preceding remarks and results. If I is a gr-maximal ideal then $\hat{I} = I\hat{R}^g$ is also a gr-maximal ideal of $\hat{R}^g$. Pick $x \in h(\hat{R}^g - \hat{I})$. There exist homogeneous $y, z \in \hat{R}^g$ with $\deg z = 0$, $z \in \hat{I}$, such that $xy = 1 - z$. The series $\Sigma_0^\infty z^n$ converges in the $\mathcal{U}_I$-gr-topology, hence it is clear that $xy \Sigma_0^\infty z^n = 1$. □

In general the $\mathcal{U}_I$-gr uniformity is not metrizable e.g. consider $R = \mathbb{Z}[X]$ with $\deg X = 1$ and let I be (p,X), then $\mathcal{U}_I$-gr is not metrizable, $\hat{R}^g = \mathbb{Z}_p[X]$, $\hat{R} = \mathbb{Z}_p[[X]]$. If we replace $\mathbb{Z}$ by a field and put $p = 0$ in the above example then we obtain an example of the metrizable case.

VI.5.9. PROPOSITION. *Let* I *be a graded ideal of the* $\mathbb{Z}$-*graded ring* R *such that the* I-*adic topology is Hausdorff (this is always the case if* R *is Noetherian). If* R *is gr*-I-*complete then* R_0 *is* I_0-*complete.*

PROOF. From $(I_0)^n \subset (I^n)_0$ it follows that $v^0(x) \leqslant v_0(x)$ for all $x \in R_0$. Consider a v^0-Cauchy sequence $\{x^{(n)}\}$ in R_0. This sequence has a limit x in the v^0-completion S_0 of R_0. Put $S = S_0 \otimes_{R_0} R$, $J = S_0 \otimes_{R_0} I$. Obviously v induces a valuation on S and v_0 extends to a valuation on S_0 and therefore the sequence $\{x^{(n)}\}$, which is also a v_0-Cauchy sequence, has a v_0-limit y in R_0. However, $\{x^{(n)}\}$ converges to x in the v^0-topology, so it must also converge to x in the v_0-topology. Consequently $x = y$ and $R_0 = S_0$ follows. □

VI.5.10. Remarks. The restriction of the $\mathcal{U}_I$-gr uniformity to R_0 need not be equal to the $\mathcal{U}_{I_0}$ uniformity. This will be the case in the situation where the center of R contains a homogeneous unit of nonzero degree. The converse of Proposition VI.5.9. need not hold and also the v_0-completeness of R_0 need not imply the gr-completeness of R. However, if R is a positively graded ring then R is gr-I-complete if and only if R_0 is I_0-complete. We refer to S. Caenepeel, F. Van Oystaeyen [18] and S. Caenepeel [17] for full detail on gr-completions.

VI.5.11. PROPOSITION. *Let* R *be a* $\mathbb{Z}$*-graded commutative ring which is gr-*I*-complete for some graded ideal* I *of* R. *Let* A *be a graded* R*-algebra which is finitely generated as an* R*-module, then* A *is gr-*AI*-complete.*

PROOF. Cf. [17]. □

Now we are ready to return to the study of the graded Brauer group.

VI.5.12. PROPOSITION. *Let* R *be a gr-local ring with gr-maximal ideal* $M = J^g(R)$. *Let* A *be a graded* R*-algebra which is finitely generated as an* R*-module. Then* A *is a separable* R*-algebra if and only if* $\overline{A} = A/MA$ *is a separable* $\overline{R} = R/M$*-algebra.*

PROOF. The "only if" part follows from the graded isomorphism of degree zero : $\overline{A} \cong A \underset{R}{\otimes} \overline{R}$. Conversely, if $\overline{A}$ is a separable $\overline{R}$-algebra then we proceed to show that A is a direct summand of $A^e = A \underset{R}{\otimes} A^0$ as left A^e-module. (We use the notations and terminology of Chapter III.2.). If $\delta : A \to J$, $a \mapsto a \otimes 1 - 1 \otimes a$ is the canonical derivation into $J = \mathrm{Ker}\, m_A$, then the separability of $\overline{A}$ yields that the induced $\overline{\delta} : \overline{A} \to \overline{J}$ is an inner derivation. Since $\overline{\delta}$ is a graded morphism of degree zero we may select $\overline{x} \in (\overline{J})_0$ such that $\overline{\delta a} = (\overline{\delta a})\overline{x}$ for all $a \in A$. Choose a representative $x \in J_0$ for $\overline{x}$. Since J is generated by $\delta(A)$ it follows from the graded version of Nakayama's lemma that $Jx = J$. The A^e-linear map $\psi : J \to J$, $a \mapsto ax$ now appears to be a graded isomorphism of degree zero which extends to $\tilde{\psi} \in \mathrm{Hom}_{A^e}(A^e,J)$ by putting $\tilde{\psi}(a) = ax$. From $(\tilde{\psi})^2 = \tilde{\psi}$ (and identifying $a - \tilde{\psi}(a)$ with $m_A(a-ax)$) it follows that $A^e \cong J \oplus A$ as graded A^e-modules. □

VI.5.13. THEOREM. *Let* R *be a gr-local ring with gr-maximal ideal* M *and suppose that* R *is gr-*M*-complete. The canonical morphism*

$$\mathrm{Br}^g(R) \to \mathrm{Br}^g(R/M)$$

is injective.

PROOF. Let A be a graded Azumaya algebra over R such that $\bar{A} = A/MA$ is trivial in $\mathrm{Br}^g(R/M)$; write $\bar{R} = R/M$. So $\bar{A} \cong M_n(\bar{R})(\underline{d})$ for some $d \in \mathbb{Z}^n$. From Proposition VI.5.9. we retain that A_0 is $(AM)_0$-complete. The matrix idempotent $\bar{e}_{11}$ has degree zero in $\bar{A}$. By the $(AM)_0$-completeness of A_0 we can lift $\bar{e}_{11}$ to an idempotent $f \in A_0$. Since Af is a finitely generated graded projective R-module, whereas R is gr-local, it follows that Af is a gr-free R-module. Representing A by left multiplication, we obtain a graded R-algebra morphism of degree zero, $\eta : A \to \mathrm{END}_R(Af)$. That η is injective is clear, surjectivity of η follows in a straightforward way from the fact that the induced morphism

$$\bar{\eta} : \overline{Af} = \bar{A}\,\bar{e}_{11} \to \overline{\mathrm{END}_R(Af)} = \mathrm{END}_{\bar{R}}(\bar{A}\,\bar{e}_{11}) = \bar{A}$$

is surjective (by the graded version of Nakayama's lemma). Therefore $A \cong \mathrm{END}_R(Af)$ is a graded isomorphism of degree zero and consequently $A \underset{gr}{\sim} 1$ in $\mathrm{Br}^g(R)$. □

By Proposition VI.5.8. the gr-completion $\hat{R}^g$ of a gr-local ring R is again gr-local, this will be essential in the following

VI.5.14. THEOREM. *Let* R *be a Noetherian gr-local ring and let* A *be a graded Azumaya algebra over* R. *There exists a gr-splitting ring* S *for* A *which satisfies all of the following conditions :*

1. S *is a faithful graded* R*-algebra*

2. S *is a gr-free* R-*module of finite rank*

3. S *is a separable graded* R-*algebra*

4. S *is gr-semilocal.*

If R *is gr*-M-*complete then* S *is gr-local.*

PROOF. Using gr-completions, the proof becomes an easy modification of Proposition 11.3. in [68]; let us provide a blue-print indicating how to deal with the minor problems of graded nature that arise. The residue algebra $\overline{A} = A \otimes_R R/M$ is a gr-c.s.a. over the gr-field $\overline{R}$ so it can be split by a Galois extension $\overline{L}$ of the form $(\overline{L})_0 \otimes_{(\overline{R})_0} \overline{R}$; we write $(\overline{R})_0 = k$, $(\overline{L})_0 = l$; see Proposition VI.3.7.
Let $l = k(\theta)$ and lift the minimal polynomial f of θ over k to a monic $F \in R[X]$ with $\deg_X F = \deg f$. Consider $R[X]$ as a graded ring by giving X degree zero ! Then $S_1 = R[X]/(F)$ is a graded R-algebra such that $\overline{S}_1 = \overline{L}$. One easily checks that S_1 is a faithful R-algebra and a gr-free R-module of finite rank. By Proposition VI.5.12., S_1 is a separable R-algebra and S_1 is gr-semilocal as one easily verifies. Moreover $\overline{L}$ splits $A \otimes_R S_1$. If R is gr-complete then so is S_1, because of Proposition VI. 5.11, and then Theorem VI.5.13. finishes the proof. In general we proceed after replacing R by S_1 i.e. we may assume that $\overline{A} \underset{gr}{\sim} 1$ in $Br^g(\overline{R})$, say $\overline{A} \cong M_n(\overline{R})(\underline{d})$ for some $\underline{d} \in \mathbb{Z}^n$.
In $(\overline{A})_0 = M_n(k)$ we consider $W_0 = k[\alpha]$, the k-separable algebra with k-basis $\{1,\alpha,\ldots,\alpha^{n-1}\}$ generated by a diagonal matrix α with n nonzero different entries. Put $W = W_0 \otimes_{(\overline{R})_0} \overline{R}$, lift α to $\beta \in A$ and let S be the R-algebra generated by $1,\beta,\ldots,\beta^{n-1}$. Exactly as in [68], 11.3. (but different from the proof of Proposition VI.5.2. !) one establishes that $\beta^n \in S$ now using the gr-completion instead of the completion. That S is separable again follows from Proposition VI.5.12. To prove that S gr-splits A one uses the double commutator theorem, cf. [10], to obtain $\mathrm{End}_S(A) = (A^0 \otimes_R S) \otimes_S A^S$, where $A^S = \{a \in A,\ sa = as \text{ for all } s \in S\}$. Now A^S is an Azumaya algebra (and graded) over S. Moreover

$$A^S = S + (A^S \cap AM) = S + A^S M,$$

so the graded version of Nakayama's lemma yields : $S = A^S$. Thus $End_S(A) = A^0 \underset{R}{\otimes} S$, i.e. S splits A and actually S gr-splits A because $End_S(A) = END_S(A)$ and the isomorphism is graded of degree zero. The other requirements for S in the statement of the theorem are easily checked. □

In order to be able to cover the whole graded Brauer group by crossed products we look for graded Galois splitting rings S such that their units are homogeneous (with an eye to graded cohomology). In order to construct the suitable normalizations of the separable splitting rings we obtained so far, we have to deal with the nontrivial idempotents of S.

VI.5.15. LEMMA. *Let* S *be a graded separable commutative extension of the gr-local domain* R *such that* S *is a projective* R*-module and a gr-splitting ring for some graded Azumaya algebra* A *over* R. *Then there exists a graded separable commutative extension* S_1 *of* R *which is also a gr-splitting ring for* A *and which is projective as an* R*-module, but which does not contain idempotent elements other than* 0 *and* 1.

PROOF. Suppose $0 \neq e \neq 1$ is idempotent in h(S). Since R is assumed to be a domain we either have $Se \cap R = 0$ or else $S(1-e) \cap R = 0$, say $Se \cap R = 0$. Obviously $S' = S/Se$ is a graded extension of R which remains projective as an R-module and separable as an R-algebra. From Proposition I.3.2., it follows that both S and S' are finitely generated R-modules, so both S and S' are gr-semilocal. It is clear that rank S' < rank S. Repeating the foregoing argument we finally end up with a graded ring S_1 containing no homogeneous idempotent elements except for 0 and 1 and satisfying the other requirements of the lemma. Consider an arbitrary idempotent f of S_1 and decompose it as $f = f_1 + \ldots + f_r$ with $\deg f_1 > \ldots > \deg f_r$. From $f^2 = f$ it follows that either $f_1^2 = 0$ or $f_1^2 = f_1$. The latter entails $f_1 = 1$, $\deg f_1 = 0$; so it will follow from $f_r^2 = 0$ or $f_r^2 = f_r$ that we either have f = 1 or else there exist nilpotent elements in S_1. So we only have to establish that S_1 is reduced (semiprime). However S_1 is a separable R-algebra and a gr-free module of finite rank over the domain R, hence unramifiedness of the minimal prime ideals of S_1 lying over zero is easily checked i.e. the intersection of the minimal prime ideals of S_1 is zero. (Actually since S_1 is gr-free the trace tr is easily calculated because it coincides with the trace of matrices in the regular representation of S_1 on the R-module S_1, so if $x \in S_1$ is nilpotent then $tr(x\_) = 0$ in $Hom_R(S,R)$ hence $x = 0$).

□

From VI.2.5. we recall that a commutative graded reduced ring S without nontrivial idempotents has all invertible elements in h(S). This leads to :

VI.5.16. THEOREM. *Let* R *be a Noetherian graded domain and let* A *be a graded Azumaya algebra over* R*, then there exists a graded Galois extension* S *of* R *with the following properties :*

1. S *gr-splits* A
2. S *is a reduced ring with only trivial idempotents.*
3. *The units of* S *are homogeneous.*

PROOF. Starting from the gr-splitting ring S_1 obtained in Theorem VI.5. 14., we may assume that S_1 satisfies condition 2 because of Lemma VI.5. 15. By the embedding theorem, [68], 3.2.9., we find an extension S_2 of S which is a normal separable extension of R and which possesses only trivial idempotent elements. As noted before, in Remark VI.5.5.1. S_2 is a graded ring and moreover, since S_1 is gr-free also S_2 will be gr-free. Since S_2 is a Galois extension of a reduced ring it is also a reduced ring, i.e. properties 2. and 3. follow. □

VI.5.17. COROLLARY. *Every graded Azumaya over a gr-local Noetherian domain* R *is gr-equivalent to a crossed product* $S[u_\sigma;\ \sigma \in G]$ *where* S *is graded Galois extension of* R *with the properties listed in Theorem VI. 5.16.* □

In the third and final part of this section we modify our foregoing results in such way that the Noetherian assumptions (i.e. also the condition $\bigcap_n M^n = 0$) may be dropped. To do this we have to replace the gr-completions by gr-Henselizations.

Let R be a gr-local commutative ring. A finitely generated graded R-algebra B is said to be gr-decomposed if B is (up to graded isomorphism of degree zero) the direct product of a finite number of gr-local rings. Recall (see e.g. [66]) that for a gr-Artinian ring R every graded prime ideal is gr-maximal; so the nilradical equals the graded Jacobson radical $J^g(R)$ hence also the Jacobson radical J(R). Consequently a gr-Artinian commutative ring will be a finite direct product of gr-Artinian gr-local rings. Since every finitely generated graded algebra over a gr-field is a gr-Artinian ring, the foregoing remarks may be used to prove :

VI.5.18. PROPOSITION. *Let* R *be a gr-local ring with unique gr-maximal ideal* M. *Every module finitely generated graded* R*-algebra* B *is semi-local and the gr-maximal ideals of* B *are just the graded prime ideals* $Q_1,\ldots,Q_n$ *lying over* M. *The natural mapping*

$$B/\bigcap_{i=1}^{n} Q_i \to \bigoplus_{i=1}^{n} Q^g_{B-Q_i}(B)/Q^g_{B-Q_i}(Q_i)$$

is a graded isomorphism of degree zero. The following conditions are equivalent :

1. B *is gr-decomposed*
2. $B \to \bigoplus_{i=1}^{n} Q^g_{B-Q_i}(B)$ *is a graded isomorphism of degree 0.*
3. *The gr-decomposition of* $\overline{B} = B/\bigcap_{i=1}^{n} Q_i$ *may be lifted to a gr-decomposition of* B.

PROOF. Formally similar to [72], Ch. I, Prop. 1,2,3. □

A gr-local commutative ring R is said to be gr-Henselian if every finitely generated graded R-algebra B is gr-decomposed. A gr-field or more general any gr-Artinian commutative ring which is gr-local is gr-Henselian.

The following result should be viewed in connection with Theorem VI. 2.3. :

VI.5.19. THEOREM. *Let* R *be a gr-local commutative ring. Then*

$$\mathrm{Id}(R) = \mathrm{Id}(R_0) = \{0,1\},$$

where Id(X) *denotes the set of idempotent elements in* X.

PROOF. We present a short proof here. First, it is clear that $\mathrm{Id}(R_0) = \{0,1\}$ because if $e \in \mathrm{Id}(R_0)$ then $R = Re \oplus R(1-e)$ cannot be gr-local. Therefore R does not contain homogeneous idempotents different from 0 and 1. If $e \in R$ is idempotent then we consider the image of e, $\bar{e}$ say, in $R/\mathrm{rad}(R) = \bar{R}$. Since $\bar{e}$ is idempotent and $\bar{R}$ is graded and reduced, it follows that $\bar{e}$ is either 0 or 1 (by the same argument as in the proof of Lemma VI.5.15.). If $\bar{e} = 1$ then $e = 1 + x$ where x is a nilpotent element, i.e. e is a unit or $e = 1$. On the other hand, $\bar{e} = 0$ would entail that e is nilpotent i.e. $e = 0$. □

VI.5.20. PROPOSITION. *Let* R *be a gr-local ring and let* B *be a graded commutative* R*-algebra which is a finitely generated* R*-module. Then, the natural map* $\mathrm{Id}(B_0) \to \mathrm{Id}(\bar{B})$ *is injective; note that* $\mathrm{Id}(\bar{B}) = \mathrm{Id}(\bar{B}_0)$. *The map is bijective exactly then when* B *is gr-decomposed.*

PROOF. $\mathrm{Id}(\bar{B}_0) = \mathrm{Id}(\bar{B})$ follows from the fact that $\bar{B}$ is a finite direct product of gr-fields. If e,e' are idempotents of degree zero such that $x = e - e' \in MB$ then

$$x^3 = (e-e')^3 = e^3 - 3e^2e' + 3e(e')^2 - (e')^3 = e - e' = x,$$

hence $x(1-x^2) = 0$. Now $x \in MB \subset J^g(B)$ and $x^2 \in J^g(B)$, thus $1 - x^2$ is a unit, so $x = 0$ follows from the foregoing (note the funny fact that the reasoning, using the third power of x, works, while the "squaring of x" does not yield anything since e and e' need not be orthogonal). Let $\overline{e}_i$ be $(0,\ldots,0,1,0,\ldots,0)$ with 1 in the ith-place considered as an element of $\overline{B} = \oplus_{i=1}^{n} Q^g_{B-Q_i}(B)/Q^g_{B-Q_i}(Q_i)$. Each idempotent of $\overline{B}$ is a sum of some of these $\overline{e}_i$, $i = 1,\ldots,n$, may be lifted to an idempotent e_i in B_0 if and only if $Q^g_{B-Q_i}(B)$ is a direct factor of B, i.e. if and only if the gr-decomposition of $\overline{B}$ can be lifted to a gr-decomposition of B. This proves the assertion. □

For the proof of the following characterization of gr-Henselian commutative rings we refer to [17].

VI.5.21. THEOREM. (S. Caenepeel). *Let* R *be a commutative gr-local ring with unique gr-maximal ideal* M*; write* $\overline{R} = R/M$. *The following statements are equivalent :*

1. R *is gr-Henselian.*
2. *Each graded* R*-algebra* B *which is a gr-free* R*-module of finite rank is a gr-decomposed algebra.*
3. *For each monic homogeneous polynomial* $F \in R[X]$, *where* $R[X]$ *is graded by putting* $\deg X = t$, *the graded* R*-algebra* $R[X]/(F)$ *is gr-decomposed, for each* $t \in \mathbb{N}$.
4. *For each monic polynomial* $F \in R_0[X]$, $R[X]/(F)$ *is a gr-decomposed* R*-algebra.*
5. *For any* $t \in \mathbb{N}$, *for each monic* $F \in R[X]$ *(where* $\deg X = t$*) such that* $\overline{F} = \overline{PQ}$ *for certain monic homogeneous polynomials* $\overline{P},\overline{Q} \in \overline{R}[X]$ *(where again* $\deg X = t$*) with* $(\overline{P},\overline{Q}) = 1$, *we have that* $F = PQ$ *are monic homogeneous polynomials lifting* $\overline{P},\overline{Q}$.
6. *For each monic* $F \in R_0[X]$ *such that* $\overline{F} = \overline{PQ}$ *for some monic* $\overline{Q},\overline{P} \in \overline{R}_0[X]$ *with* $(\overline{Q},\overline{P}) = 1$, *we have* $F = PQ$, *where* P,Q *are monic polynomials in* $R_0[X]$ *lifting* P *and* Q.
7. R_0 *is Henselian.*
8. *A monic polynomial* $F \in R_0[X]$ *such that* $\overline{F} \in R_0[X]$ *has a simple zero* $\overline{a}$ *in* $\overline{R}_0$ *has a zero* a *in* R_0 *lifting* $\overline{a}$.
9. *A monic homogeneous* $F \in R[X]$, *where* $\deg X = t$, *such that* $\overline{F} \in \overline{R}[X]$,

where $\deg X = t$, *has a simple zero* $\bar{a}$ *in* $\bar{R}$, *has a zero* a *in* R *lifting* $\bar{a}$. □

VI.5.22. Remark. It is clear that the technique of gr-Henselization is related to constructions in degree zero in a much better way than the technique of gr-completion because of the equivalence 1 ⇔ 7. On the other hand every gr-complete gr-local ring is gr-Henselian, as one easily verifies.

Returning to the study of the graded Brauer group we first note that the following generalization of Theorem VI.5.13. holds :

VI.5.23. THEOREM. *Let* R *be a gr-local gr-Henselian ring, then the natural homomorphism* $\mathrm{Br}^g(R) \to \mathrm{Br}^g(R/M)$ *is injective. Moreover, for every graded extension* S *of* R *which is finitely generated as an* R*-module, we have that* $\mathrm{Br}^g(S) \to \mathrm{Br}^g(\bar{S})$ *where* $\bar{S} = S/MS$, *is an injective homomorphism.*

PROOF. Similar to 2.2. in [68]. □

We introduce the following terminology. In general, a graded Azumaya algebra A over a graded ring R which is gr-split by a graded extension S of R gives rise to a graded isomorphism of degree zero

$$\sigma : A \underset{R}{\otimes} S \to \mathrm{END}_S P,$$

for some graded faithfully projective S-module P. Now consider the graded isomorphism φ of degree zero defined by the following commutative diagram :

$$\begin{array}{ccc} S \underset{R}{\otimes} A \underset{R}{\otimes} S & \xrightarrow{1\otimes\sigma} & \mathrm{END}_{S^{(2)}} P_1 \\ \Big\downarrow {\scriptstyle t\otimes 1} & & \Big\downarrow {\scriptstyle \varphi} \\ A \underset{R}{\otimes} S \otimes S & \xrightarrow{\sigma\otimes 1} & \mathrm{END}_{S^{(2)}} P_2 \end{array}$$

where t is the switch map, where P_1, P_2 are graded faithfully projective $S^{(2)}$-modules (further notation stems from Section VI.1.).

This morphism φ is induced by an $S^{(2)}$-isomorphism of degree zero $f : P_1 \underset{S^{(2)}}{\otimes} I \to P_2$ for some graded $S^{(2)}$-progenerator I of rank one. If $[I] = 1$ in $\mathrm{Pic}^g(S^{(2)})$ then we may say that (S,P,σ,φ,f,I) determines a <u>good gr-splitting</u> for A. If $[I] = 1$ in $\mathrm{Pic}_g(S^{(2)})$ then (S,P,σ,φ,f,I) is said to be a <u>very good gr-splitting</u> for A. It is easily seen that a graded faithfully flat extension S' of S will also determine a (very) gr-splitting for A if S does determine a (very) good gr-splitting for A.

We now strengthen Proposition VI.3.7. as follows :

VI.5.24. <u>THEOREM</u>. *Let* R *be a commutative gr-field and consider a graded Azumaya algebra* A *over* R. *There exists a Galois field extension* l *of* $k = R_0$ *such that* $S = R \underset{R_0}{\otimes} l$ *determines a good gr-splitting for* A.

1. *If* R *is trivially graded i.e.* $R = R_0 = k$ *then* S *determines a very good gr-splitting for* A.
2. *If* R *is not trivially graded for* $R = k[T,T^{-1}]$, $S = l[T,T^{-1}]$, *and* $S' = l[T^{\frac{1}{\alpha}},T^{-\frac{1}{\alpha}}]$ *where* $d = \deg T$ *determines a very good gr-splitting of* A *(which is also a graded faithfully flat extension).*

<u>PROOF</u>. 1. If $R = k$ then A is a c.s.a. equipped with a gradation (e.g. an example of a gradation on $M_n(k)$ is given in [66]). If l is a Galois extension of k splitting A then $\mathrm{Pic}_g(l) = \mathbb{Z}$. It is easy to see that a graded $l \underset{R}{\otimes} l$-progenerator I has the property $[I] = 1$ in $\mathrm{Pic}_g(l)$ and thus it is isomorphic in degree zero to $J_1 \underset{l}{\otimes} J_2^{-1}$ for certain graded l-progenerators J_1,J_2 of rank one. The statement is now easily verified.

2. From Proposition VI.3.7. we know that there is a gr-field $S = l[T,T^{-1}]$ where l is a Galois extension of k, such that A is gr-split by S. Put $\deg T = d$. First, since $S \underset{k}{\otimes} S$ is a gr-semilocal ring, $\mathrm{Pic}^g(S \underset{k}{\otimes} S) = 1$, so S certainly determines a good gr-splitting for A. Obviously, S is a graded faithfully flat extension of S, hence S' gr-splits A. Now $S' \underset{k}{\otimes} S'$ is again gr-semilocal but it contains a homogeneous unit of degree one,

therefore $Pic_g(S' \otimes S') = 1$. Hence, S' determines a very good gr-splitting for A. Note however that S' need not be a separable extension. □

VI.5.25. PROPOSITION. *Let* R *be a gr-local ring and let* A *be a graded Azumaya algebra over* R. *There exists a good gr-splitting ring* S *for* A, *such that* S *is a separable* R*-algebra (and a graded étale covering of* R*) which is a gr-free* R*-module of finite rank and such that* S *is gr-semilocal.*

PROOF. Replace the gr-completion by the gr-Henselization in the proof of Theorem VI.5.14. and use : VI.5.23. instead of VI.5.13., VI.5.24. instead of VI.3.7. □

VI.5.26. THEOREM. *Let* R *be a gr-local ring and let* A *be a graded Azumaya algebra over* R. *Then there exists a graded Galois extension* S *of* R *which is a gr-free* R*-module of finite rank, gr-semilocal, and which determines a good gr-splitting for* A.

PROOF. Let S be the graded separable extension of R obtained in Proposition VI.5.25. By Theorem VI.5.19., R does not contain idempotent elements different from 0 and 1. From Villamayor, Zelinski [102], Proposition 1.2., Lemma 2.3. bis, it follows that $S \cong Se_1 \oplus \ldots \oplus Se_n$, where the e_i are idempotents of degree zero and each Se_i is a graded commutative separable R-algebra and a finitely generated projective R-module. Clearly each Se_i gr-splits A. Normalizing Se_1 say, yields a Galois extension with the desired properties. □

Finally, we have established that for a considerable class of rings (gr-local rings) we may describe the elements of the graded Brauer group by crossed products. Indeed, we obtained good gr-splitting rings which are gr-semilocal and Galois extensions of R possessing no nontrivial units. Clearly this situation invites a graded-cohomology theory. This will be developed, in a more general setting, in the following Section VI.6.

VI.6. Cohomological Interpretation.

Using Amitsur cohomology we introduce graded-cohomology groups and use these to describe $Br^g(S/R)$, the part of the graded Brauer group which may be split by the graded extension S of R. We apply this theory to obtain Galois cohomological results in the gr-local case and further to relate $Br^g(R)$ and $Br(R)$ in some particular cases. For general cohomology theory we refer to [19], in particular we assume familiarity with the basic concepts such as complexes and mappings between them.

Let C be a general commutative ring and let (A,Δ) and (B,δ) be two C-complexes. Consider a mapping $d : A \to B$ of complexes of degree zero, i.e. :

$$\begin{array}{ccccccccc} \cdots & \longrightarrow & A^{n-1} & \xrightarrow{\Delta_{n-1}} & A^n & \xrightarrow{\Delta_n} & A^{n+1} & \longrightarrow & \cdots \\ & & \downarrow{\scriptstyle d^{n-1}} & & \downarrow{\scriptstyle d^n} & & \downarrow{\scriptstyle d^{n+1}} & & \\ \cdots & \longrightarrow & B^{n-1} & \xrightarrow[\delta_{n-1}]{} & B^n & \xrightarrow[\delta_n]{} & B^{n+1} & \longrightarrow & \cdots \end{array}$$

with $\Delta_n\Delta_{n-1} = 0$, $\delta_n\delta_{n-1} = 0$, $\delta_{n-1}d^{n-1} = d^n\Delta_{n-1}$ for all $n \in \mathbb{Z}$. As in [19], VI.3., we define :

$$H^n(A) = \ker \Delta_n/\mathrm{Im}\ \Delta_{n-1},\ H^n(B) = \mathrm{Ker}\ \delta_n/\mathrm{Im}\ \delta_{n-1}.$$

Starting from A and B we define a new complex (D,∇) by :

$$D^n = A^n \times B^{n-1},\ \nabla_n : D^n \to D^{n+1}$$

given by $\nabla_n(a,b) = (\nabla_n a,\ \delta_{n-1}b - d^n a)$ i.e. :

$$\cdots \longrightarrow A^{n-1} \times B^{n-2} \xrightarrow[\nabla_{n-1}]{} A^n \times B^{n-1} \xrightarrow[\nabla_n]{} A^{n+1} \times B^n \longrightarrow \cdots$$

We define the n-th B-cohomology group of A by $H^n_B(A) = \mathrm{Ker}\ \nabla_n/\mathrm{Im}\ \nabla_{n-1}$. With these notations we have :

VI.6.1. PROPOSITION. *The following sequence is exact*

$$\cdots \to H^{n-1}(B) \to H^n_B(A) \to H^n(A) \to H^n(B) \to H^{n+1}_B(A) \to \cdots$$

PROOF. The exact sequence : $0 \to B^{n-1} \to D^n \to A^n \to 0$ induce an exact sequence of complexes, so we may apply [19], VI.3. □

Let R be a graded commutative ring, then recall from Proposition VI. 2.1., that the morphism $d_0 : U(R) \to G(R)$ fits in a long exact sequence :

$$1 \to U_0(R) \to U(R) \xrightarrow[d_0]{} G(R) \xrightarrow[\psi]{} \mathrm{Pic}_g(R) \xrightarrow[\varphi]{} \mathrm{Pic}^g(R) \to 1.$$

We use the notation introduced in Section VI.1., before VI.1.3. It is clear that we may derive from d_0 a mapping d between the Amitsur complexes :

$$\begin{array}{ccccccccc} 1 & \longrightarrow & U(S) & \xrightarrow{\Delta_0} & U(S^{(2)}) & \xrightarrow{\Delta_1} & U(S^{(3)}) & \longrightarrow & \cdots \\ & & \downarrow d_0 & & \downarrow d_1 & & \downarrow d_2 & & \\ 1 & \longrightarrow & G(S) & \xrightarrow[\delta_0]{} & G(S^{(2)}) & \xrightarrow[\delta_1]{} & G(S^{(3)}) & \longrightarrow & \cdots \end{array}$$

where S is a commutative graded R-algebra. The foregoing general construction of $H^n_B(A)$, taking for A the Amitsur complex C(S/R,U) of U, and for B the Amitsur complex C(S/R,G) of G, yields the graded-cohomology group of U which we denote by $H^n_B(A) = H^n_{gr}(S/R,U)$.
Another interesting situation where we apply the general construction of $H^n_B(A)$ succesfully is the following. Let S be a graded Galois extension of R with Galois group G and let K(G,U(S)) and K(G,G(S)) be the complexes associated to the G-$\mathbb{Z}$-modules U(S) and G(S). In this case, replacing A and B in the definition of $H^n_B(A)$ by the foregoing complexes we obtain $H^n_B(A) = H^n_{gr}(G,U(S))$.
As a direct consequence of [56], Proposition V.1.6. we have an isomorphism $H^n_{gr}(G,U(S)) \cong H^n_{gr}(S/R,U)$. In terms of Amitsur cohomology, the exact sequence of Proposition VI.6.1. may now be rewritten as

$$1 \to H^0_{gr}(S/R,U) \to H^0(S/R,U) \to H^0(S/R,G) \to H^1_{gr}(S/R,U) \to$$

$$\to H^1(S/R,U) \to H^1(S/R,G) \to H^2_{gr}(S/R,U) \to \cdots$$

To fix notations : R is a commutative $\mathbb{Z}$-graded ring and S is a graded R-algebra which is also a graded R-progenerator, we will write $H^n(U)$,

$H^n_{gr}(U),\ldots$, for $H^n(S/R,U)$, $H^n_{gr}(S/R,U),\ldots$. With these notations and conventions we have :

VI.6.2. THEOREM. 1. $H^0(U_0) = U_0(R)$.
2. $H^1(U_0) = \mathrm{Pic}_g(S/R) := \mathrm{Ker}(\mathrm{Pic}_g(R) \to \mathrm{Pic}_g(S))$.
3. *The following sequence is exact :*

$$1 \to H^1(U_0) \xrightarrow{\alpha} \mathrm{Pic}_g(R) \xrightarrow{\beta} H^0(\mathrm{Pic}_g) \xrightarrow{\gamma} H^2(U_0) \to$$
$$\xrightarrow[\delta]{} \mathrm{Br}^g(S/R) \xrightarrow[\xi]{} H^1(\mathrm{Pic}_g) \xrightarrow[\eta]{} H^3(U_0)$$

PROOF. 1. Follows from the theorem of faithfully glat descent of elements, cf. [56], II.2.1.

2. and 3. are straightforward graded versions of the classical Chase-Rosenberg sequence as given in [56] but taken into account the graded versions of the classical results recalled here in Section VI.1. and 2. Let us limit ourselves here to the description of the connecting homomorphisms.

α. If $t \in U_0(S^{(2)})$ is a **co-cycle** in degree zero then multiplication by t is a descent datum in degree zero : $S^{(2)} \to S^{(2)}$, defining an invertible graded module I. Put $\alpha(t) = [I]$.

β. The morphism β is just extension of scalars.

γ. If $[I] \in H^0(\mathrm{Pic}_g)$, there exists an $S^{(2)}$-isomorphism $\varphi : I_1 \to I_2$ which is graded of degree zero. Now $\varphi_2^{-1}\varphi_3\varphi_1$ is multiplication by a unit $u \in U_0(S^{(3)})$, which is a cocycle. Define $\gamma([I]) = [u]$.

δ. Let P be the S-module $S \otimes_R S$ where S acts on the first factor. A cocycle $t \in U_0(S^{(3)})$ defines an $S^{(2)}$-isomorphism $f(t) : P_1 \to P_2$ which is defined by multiplication by t and switching the second and third factor. Clearly, f(t) induces a descent datum in degree zero,

$$\varphi(t) : S \otimes_R \mathrm{END}_S(P) \to \mathrm{END}_S(P) \otimes_R S.$$

Therefore it defines a graded Azumaya algebra A(t) over R,

$$A(t) = \{x \in \mathrm{END}_S(P);\ \varphi(t)\varepsilon_1 x = \varepsilon_2 x\}.$$

We put $S(t) = [A]$.

ξ. Given $[A] \in Br^g(S/R)$, this yields a graded splitting of A by S, say $\tau : S \otimes_R A \to END_S(Q)$ where τ is a graded isomorphism of degree zero and Q is a graded projective S-module. Let $\varphi = \tau_3\tau_1\tau_2^{-1}$

$$\varphi : \tau_3\tau_1\tau_2^{-1} : END_{S^{(2)}}(Q_1) \to END_{S^{(2)}}(Q_2),$$

where φ is induced by some $f : Q_1 \otimes_{S^{(2)}} I \to Q_2$, where $I \in H^1(Pic_g)$. Put $\xi([A]) = [I]$.

η. For $[I] \in H^1(Pic_g)$ there exists an $S^{(3)}$-isomorphism of degree zero $f : I_1 \otimes_{S^{(3)}} I_3 \to I_2$. One verifies that $f_4^{-1}f_2^{-1}f_3f_1$ is just multiplication by some unit $u \in U_0(S^{(4)})$, which is a co-cycle. We put $\eta([I]) = [u]$. □

VI.6.3. Remark. For 1. and 2. in the above theorem it would be sufficient to have that S is a graded faithfully flat R-algebra.

VI.6.4. PROPOSITION. *Suppose that* S *is a graded faithfully flat* R-*algebra, then :*

$$Im(H^1_{gr}(U) \to H^1(U)) = Ker(H^1(U) \to H^1(G)) = Pic^g(S/R).$$

PROOF. The first identity follows by Proposition VI.6.1. Note that $Pic^g(S/R) = Ker(Pic^g(R) \to Pic^g(S))$. Now consider $[t] \in ker(d_1)$. There is a $T \in G(S)$ such that $d_1(t) = T_1 \otimes_{S^{(2)}} T_2^{-1}$. Consider the map $: S^{(2)} \to d_1(t)$ given by multiplication by t. This φ is a graded isomorphism of degree zero! Tensoring by T_2 we obtain an $S^{(2)}$-isomorphism which is again graded of degree zero, $: T_2 \to T_1$, and which is still described by multiplication by t. Since $t_2 = t_3t_1$ we have $\varphi_2 = \varphi_3\varphi_1$ and therefore is a descent datum for T in degree zero, defining an $[I] \in Pic_g(R)$ such that $S \otimes_R I$ and S are graded isomorphic in degree zero. Put $\alpha[t] = \varphi(I)$ (see Proposition VI.2.1), then $\alpha[t]$ $Pic^g(R) \subset Pic(R)$. In fact α is the restriction of the classical morphism : $\alpha : H^1(U) \to Pic(R)$ studied in [56];

well-defined and Im $\alpha \subset \mathrm{Pic}^g(S/R)$. Moreover α is injective. Actually Im $\alpha = \mathrm{Pic}^g(S/R)$. Indeed if $[\underline{I}] \in \mathrm{Pic}^g(S/R)$ let $I \in \varphi^{-1}([\underline{I}])$, where φ is the morphism of Proposition VI.2.1., $\varphi : \mathrm{Pic}_g(R) \to \mathrm{Pic}^g(R)$. Let $\underline{f} : S \underset{R}{\otimes} I \to S$ be an S-isomorphism (not necessarily of degree zero !). Put $T = G(\underline{f})$. Then we have that $f : S \underset{R}{\otimes} I \to T$ is a graded S-isomorphism of degree zero (see Lemma VI.2.2.). Define φ by the commutativity of the following diagram of graded $S^{(2)}$-isomorphisms of degree zero :

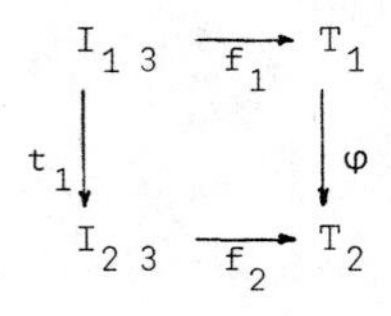

(Notation of Section VI.1., before VI.1.3. !)
So φ must be multiplication by a unit $t \in U(S^{(2)})$, cf.[56], I.6.2., and t is a cocycle. Since $d_1(t) = T_1 \underset{S^{(2)}}{\otimes} T_2^{-1}$, $t \in \mathrm{Ker}(d_1)$ follows. This proves the second equality. □

VI.6.5. <u>THEOREM</u>. (S. Caenepeel). *With notations and conventions as before, we obtain the following long exact sequence :*

$$1 \to U_0(R) = H^0(U_0) = H^0_{gr}(U) \to U(R) = H^0(U) \to H^0(G) \to H^1_{gr}(U) \to$$
$$\mathrm{Pic}^g(R) \to H^0(\mathrm{Pic}^g) \to H^2_{gr}(U) \to \mathrm{Br}^g(S/R) \to H^1(\mathrm{Pic}^g) \to H^3_{gr}(U).$$

<u>PROOF</u>. Consists of checking exactness at each step. We give the complete proof in the appendix to this section, but prefer now to go on with some of its consequences. □

VI.6.6. <u>COROLLARY</u>. (Crossed Product Theorems). *If* S *is a graded faithfully flat commutative* R*-algebra then :*

1. $\mathrm{Pic}^g(S) = \mathrm{Pic}^g(S \underset{R}{\otimes} S) = 1$, *implies that :* $\mathrm{Br}^g(S/R) = H^2_{gr}(S/R,U)$.
2. $\mathrm{Pic}(S) = \mathrm{Pic}_g(S \otimes S) = 1$, *implies that :* $\mathrm{Br}^g(S/R) = H^2(S/R,U_0)$.

<u>PROOF</u>. Direct from Theorem VI.6.5. □

If S is a graded Galois extension of R then all cohomology groups $H^n(F) = H^n(S/R,F)$ may be replaced by Galois-cohomological groups, $H^n(G,F(S))$, for the functor F. For example we may apply these techniques to the case of a gr-field $R = k[T,T^{-1}]$ and obtain the following extended version of Theorem VI.3.8. :

VI.6.7. <u>THEOREM</u>. *Let* R *be the gr-field* $k[T,T^{-1}]$ *where* deg T = d. *Every graded Azumaya algebra over* R *may be represented by a crossed product. Moreover* $Br^g(R) = Br(k) \oplus H^2_{gr}(Gal(\bar{k}/k),d\mathbb{Z})$, *where* $\bar{k}$ *is the algebraic closure of* k.
If char(k) *does not divide* d *then every graded Azumaya algebra over* R *can be represented by a crossed product in degree zero; in this case :*

$$Br^g(R) = Br(k) \oplus H^2(\mathbb{Z}/d\mathbb{Z},U(\bar{k})).$$

If deg T = 1, $Br^g(R) = Br(k)$.

<u>PROOF</u>. By Proposition VI.3.7., a graded Azumaya algebra A over R has a gr-splitting ring $l[T,T^{-1}]$ where l/k is a Galois extension with $G = Gal(l/k)$. Since $Pic^g(S) = 1$,

$$Br^g(S/R) = H^2_{gr}(G,U(S) = H^2_{gr}(G,U(l) \oplus \langle T\rangle) = H^2(G,U(l)) \oplus H^2_{gr}(G,\langle T\rangle)$$

(where $\langle T\rangle$ is the cyclic group generated by T), and

$$\begin{aligned} Br^g(R) &= \varinjlim H^2(Gal(l/k),U(l)) \oplus \varinjlim H^2_{gr}(Gal(l/k),\langle T\rangle) \\ &= Br(k) + H^2_{gr}(Gal(\bar{k}/k),d\mathbb{Z}) \end{aligned}$$

where we identify $\langle T\rangle$ with $d\mathbb{Z}$, $G(\bar{k}[T,T^{-1}])$ with $\mathbb{Z}$, and where the limit is over all Galois extensions l of k. Actually $H^2_{gr}(Gal(\bar{k}/k),d\mathbb{Z})$ is the kernel of the mapping $H^2(Gal(\bar{k}/k),d\mathbb{Z}) \to H^2(Gal(\bar{k}/k),\mathbb{Z})$. If char(k) $\nmid$ d, then we may replace l by the Galois extension L obtained by splitting the polynomial $Y^d = 1$ i.e. we may take L to be a cyclotomic extension of l. Then replace S by $S' = L[X,X^{-1}]$ where $X^d = T$ and define the gradation of S' by putting deg X = 1. If $1,\omega,\ldots,\omega^{d-1}$ are the d-th roots of 1 in L then $Gal(S'/S) = \{1,\tau_1,\ldots,\tau_{d-1}\}$ where each τ_i

is determined by putting $\tau_i(X) = \omega^i X$. Since $\mathrm{Pic}_g(S') = 1$ (there are units of degree one !) it follows that $\mathrm{Br}^g(S'/r) = H^2(\mathrm{Gal}(S'/R),U(L))$. Taking limits over such S'/R, it is evident that

$$\mathrm{Br}^g(R) = \mathrm{Br}(k) \oplus H^2(\mathbb{Z}/d\mathbb{Z},U(\bar{k})).$$

The last statement of the theorem is now obvious (it had been proved before by equally easy methods). □

Now we return to the situation where R is a gr-local ring and where S is a graded Galois extension of R, with Galois group G say. There is a canonical morphism :

$$\deg_S^2 : H^2(G,U(S)) \to H^2(G,\mathbb{Z}),$$

given by $\deg^2(c)(\sigma,\tau) = \deg c(\sigma,\tau)$, for $\sigma,\tau \in G$. We write $\deg^2$ for the map induced on $\varinjlim_G H^2(G,U(S))$ by the maps $\deg_S^2$, where the limit is taken over the Galois extensions S of R. By Theorem VI.5.26. we only have to consider Galois extensions S of R which are gr-free R-modules of finite rank, gr-semilocal and reduced; consequently the units of such an S are homogeneous.

VI.6.8. THEOREM. *Let* R *be a gr-local domain, then* $\mathrm{Br}^g(R) = \mathrm{Ker}\,\deg^2$.

PROOF. We only have to show that for a gr-splitting ring S which is a graded Galois extension of R, we have $\mathrm{Br}^g(S/R) = \mathrm{Ker}\,\deg_S^2$.
By Corollary VI.6.6.1. we only have to show that $H^2_{gr}(G,U(S)) = \mathrm{Ker}\,\deg_S^2$. First consider $\alpha \in \mathrm{Ker}\,\deg_S^2$, i.e. such an α may be represented as a crossed product $A = S[u_\sigma; c(\sigma,\tau)]$ where $\deg c(\sigma,\tau) = d_\sigma + d_\tau - d_{\sigma\tau}$ for some $d_\sigma \in \mathbb{Z}$, for $\sigma,\tau \in G$. It is clear that we may define a gradation on A by putting $\deg u_\sigma = d_\sigma$ for all $\sigma \in G$ and in this way A becomes a graded R-algebra representing an element of $H^2_{gr}(G,U(S))$. Conversely, consider a graded Azumaya algebra over R, A say, which represents an element $\alpha \in H^2(G,U(S))$ i.e. we may assume $A = S[u_\sigma; c(\sigma,\tau)]$ for some cocycle c representing α. We claim that u_σ is homogeneous, for each $\sigma \in G$. If

this holds then the cocycle condition $u_\sigma u_\tau = c(\sigma,\tau)u_{\sigma\tau}$ entails the relation : $\deg c(\sigma,\tau) = \deg u_\sigma + \deg u_\tau - \deg u_{\sigma\tau}$ or $\deg^2 c$ is trivial in $H^2(G,\mathbb{Z})$, as described; so let us establish the claim. Select a fixed σ, for the moment, and write $u = u_\sigma$, $v = u_\sigma^{-1}$; moreover we may assume that $u_1 = 1$. Write $u = u_1 +\ldots+ u_n$, $v = v_1 +\ldots+ v_n$ with $\deg u_i = d_i = -\deg v_i$ and $d_n >\ldots> d_1$ (this is possible if we allow some of the u_i, v_i to be zero). If $u \notin h(A)$ then $d_n \neq d_1$ and from $1 = uv : vu$ it follows that :

$$0 = (uv)_{d_n-d_1} = (vu)_{d_n-d_1} = u_n v_1 = v_1 u_n.$$

Suppose, inductively, that we have already established $0 = u_n v_j = v_j u_n$ for $j = 1,\ldots,i-1$, with $i < n$, hence $d_i < d_n$. Then we obtain the relation :

$$0 = (1)_{d_n-d_1} = u_n v_i + \sum_{\substack{d_n-d_i=d_l-d_k \\ l\neq k}} u_l v_k.$$

From $d_n - d_l > 0$ we may deduce that $d_i - d_k > 0$ i.e. the non-zero contributions in the sum appear with $k < i$. Multiply the above relation by $u_n v_i$; we then obtain $(u_n v_i)^2 = 0$. The structure of A is defined by $u_\sigma s = \sigma(s)u_\sigma$ for all $s \in S$, all $\sigma \in G$; so restricting to homogeneous $s \in S$ we obtain : $us = \sigma(s)u$ and $u_j s = \sigma(s)u_j$ for all $j = 1,\ldots,n$. On the other hand, the cocycle condition (combined with the assumption $u_1 = 1$) yields $u_{\sigma^{-1}} = c(\sigma,\sigma^{-1})v$. Therefore $vs = \sigma^{-1}(s)v$ for all $s \in S$ and by restricting to homogeneous $s \in S$: $v_j s = \sigma^{-1}(s)v_j$ for all $j = 1,\ldots,n$. This shows that any product $u_i v_j$ with $i,j \in \{1,\ldots,n\}$ commutes with the whole of S. Hence $u_i v_j \in S$, as S is a maximal commutative subring. But then the fact that S is semiprime (it is a Galois extension of a domain) implies that $u_n v_i = 0$. A symmetric argument yields that $v_i u_n = 0$ too. So far we already established that $u_n v_i = v_i u_n = 0$ for $i = 1,\ldots,n-1$.

Another symmetric argumentation, starting off at the lowest degree terms, yields $u_i v_n = v_n u_i = 0$ for all $i = 1,\ldots,n-1$. The relation is then reduced to : $1 = (uv)_0 = u_1 v_1 +\ldots+ u_n v_n$. Multiplication by $u_n v_n$ yields $u_n v_n = (u_n v_n)^2$. But since $u_n v_n \in S$ and S does not contain idempotents except 0 and 1, it follows that either $u_n v_n = 1$ or $u_n v_n = 0$ (note : we used the remark preceding the theorem i.e. by Theorem VI.5.26. we were allowed to assume that S has all the properties mentioned there). If $u_n v_n = 0$ then $uv = 1$ reduces to : $(u_1 + \ldots + u_{n-1})(v_1 + \ldots + v_{n-1}) = 1$, because $0 = u_n v_i = u_n v_n = u_j v_n$ for $i,j \in \{1,\ldots,n\}$. So repetition of the foregoing argument eventually leads to the situation where $u_j v_j = 1$ for some $j \in \{1,\ldots,n\}$. Since $u^{-1}u_j$ commutes with S it follows that $u_j = su$ for some $s \in S$. Then $suv_j = 1$ with $uv_j \in S$ entails that $s \in U(S)$ hence s is homogeneous. Consequently $u = s^{-1}u_j$ is then homogeneous too. Since we have fixed $u = u_\sigma$ arbitrary, it follows that all u_σ, $\sigma \in G$, are homogeneous in A and our claim has been established. □

VI.6.9. COROLLARY. *If* R *is a gr-local ring and* S *is a graded Galois extension of* R *with the properties mentioned in Theorem VI.5.26. then we have an exact sequence of group homomorphisms*

$$1 \to H^2_{gr}(G,U(S)) \to H^2(G,U(S)) \xrightarrow{\deg^2} H^2(G,\mathbb{Z}) \to M \to 1$$

where M *is an* e*-torsion group and* e *is the smallest* $e \in \mathbb{N}$ *such that* R_e *contains a unit.*

PROOF. If $d \in H^2(G,\mathbb{Z})$ then ed is a 2-cocycle such that $(ed)(\sigma,\tau) \in \mathbb{Z}$ for all $\sigma,\tau \in G$. Put $c(\sigma,\tau) = T^{d(\sigma,\tau)}$ where T is the unit of degree e in R. Then $c : G \times G \to U(S)$ defines an element of $H^2(G,U(S))$ such that $\deg^2 c = ed$. Therefore $H^2(G,\mathbb{Z})/\mathrm{Im}\,\deg^2$ is e-torsion. Exactness at the other places in the sequence follows from the theorem. □

For the final theorem of this section and its corollaries we do not strive for full generality. As a matter of fact similar result could be derived for gr-Dedekind rings, or by weakening the conditions on S a little. We do not go into this here because our aims do not require full generality. The conventions from here on are that R is a gr-local domain, S is a graded Galois extension of R with the properties mentioned in Theorem VI.5.26., and moreover we assume that $S = S_0 \otimes_{R_0} R$ for some Galois extension S_0 of R_0 with Galois group G (this is the point where weaker hypotheses might be used). Let $e \in \mathbb{N}$ be the smallest positive integer such that R_e contains a unit, T say. With these conventions we have the following :

VI.6.10. <u>THEOREM</u>. 1. $Br(S/R)/Br(S_0/R_0) \oplus H^2(G,\mathbb{Z})$ *is an e-torsion group.* 2. $Br^g(S/R)/Br(S_0/R_0)$ *is an e-torsion group.*

<u>PROOF</u>. 1. Foregoing results allow to identify : $Br(S/R) = H^2(G,U(S))$, $Br^g(S/R) = H^2_{gr}(G,U(S))$. If $c \in H^2(G,U(S))$ then $c_0 = (ec)T^{-d}$, where $d = \deg^2 c$, represents an element of $H^2(G,U(S_0))$. Consequently $ec = c_0 T^d$ may be viewed as an element of $H^2(G,U(S_0)) \oplus H^2(G,\mathbb{Z})$.

2. If $c \in H^2_g(G,U(S))$ then $c \in \operatorname{Ker} \deg^2$ by Theorem VI.6.8. Write $c(\sigma,\tau) = d_\sigma + d_\tau - d_{\sigma\tau}$ for certain $d_\sigma \in \mathbb{Z}$, $\sigma,\tau \in G$. As in 1. we obtain $(ec) = T^d c_0$ but this means that ec and c_0 are representing the same element of $Br^g(S/R)$, because $T^{d(\sigma,\tau)} = T^{d_\sigma}.T^{d_\tau}.T^{-d_{\sigma\tau}}$ for all $\sigma,\tau \in G$, i.e. is a trivial cocycle. □

VI.6.11. <u>COROLLARIES</u>

1. *If* e = 1 *then* $Br(R/S) = Br(S_0/R_0) \oplus H^2(G,\mathbb{Z})$ *and* $Br^g(S/R) = Br(S_0/R_0)$.

2. *If* R *is of the form* $R = R_0[X,X^{-1}]$ *with* deg X = e *then we may change the gradation of* R *by giving* deg X *any other value in* $\mathbb{Z}$. *Hence :* $Br(S/R)/Br(S_0/R_0) \oplus H^2(G,\mathbb{Z})$ *is e-torsion for any e, therefore* $Br(S/R) = Br(S_0/R_0) \oplus H^2(G,\mathbb{Z})$. *Although* $Br^g(S/R)/Br(S_0/R_0)$ *is e-torsion, the fact that* $Br^g(S/R)$ *varies if we change the gradation of* R *does not allow to conclude that* $Br^g(S/R)$ *is in any sense "very close" to* $Br(S_0/R_0)$.

3. *If* R *is a discrete gr-valuation ring, then the results of Section VI.5. and the foregoing results imply that any graded Azumaya algebra over* R *has a gr-splitting ring with all the properties listed above. So the foregoing results now may be combined in :*

a. $Br^g(R) = \varinjlim_G H^2_{gr}(G,U(S))$

b. $\varinjlim_G (H^2(G,U(S))/H^2_{gr}(G,U(S)))$ *is an e-torsion group.*

c. $Br^g(R)/Br(R_0)$ *is an e-torsion group.* □

VI.6.12. THEOREM. *Let* R *be a gr-Dedekind ring and let* S *be a gr-Dedekind ring which is a Galois extension of* R *with Galois group* G.
Let $S_{(n)}$ *be the graded ring obtained by equiping the ungraded ring* S *with the gradation defined by* $(S_{(n)})_{np} = S_p$ *for* $p \in \mathbb{Z}$. *We identify the group* $H^2_{gr}(G,U(S_{(n)}))$ *with its image in* $H^2(G,U(S))$, *(by forgetting gradation). Then we have* $H^2(G,U(S)) = \bigcup_{n\in\mathbb{Z}} H^2_{gr}(G,U(S_{(n)}))$.

PROOF. Pick $c \in H^2(G,U(S))$. Then $d = \deg^2 c \in H^2(G,\mathbb{Z})$ has finite order, n say. So $(nd)(\sigma,\tau) = d_\sigma + d_\tau - d_{\sigma\tau}$ for all $\sigma,\tau \in G$ for some elements $d_\sigma \in \mathbb{Z}$, $\sigma \in G$. In the graded structure of S, $c(\sigma,\tau)$ has degree $d(\sigma,\tau)$, but in the graded structure of $S_{(n)}$, $\deg_{(n)} c(\sigma,\tau) = nd(\sigma,\tau)$. In other words $\deg^2_{(n)} c = nd$, but this states that $c \in \operatorname{Ker} \deg^2_{(n)}$, or c represents an element of $H^2_{gr}(G,U(S_{(n)}))$. □

VI.6.13. COROLLARIES.
1. *If* $R = R_0[X,X^{-1}]$ *where* R_0 *is a discrete valuation ring with perfect residue field* k, *then* $Br(R) = \bigcup_{n\in\mathbb{Z}} Br^g(R_{(n)})$.
2. *(Special case). If* $R = k[T,T^{-1}]$ *where* k *is a perfect field then any Azumaya algebra* A *over* R *is of the form* $M_n(\Delta[X,X^{-1},\varphi])$, *where* $n \in \mathbb{N}$, Δ *is a skewfield with center* $Z(\Delta)$, φ *is an automorphism of* Δ *such that* $Z(\Delta)^\varphi = k$ *and some power of* φ *becomes an inner automorphism of* Δ.

PROOF. 2. (1 is similar !). By Tsen's theorem any Azumaya algebra A over $k[T,T^{-1}]$ may be split by a Galois extension l/k. Therefore we may take direct limits on both sides in the equality

$$H^2(G,U(S)) = \bigcup_{n\in\mathbb{Z}} H^2_{gr}(G,U(S_{(n)})),$$

established in Theorem VI.6.12. The result then follows from the description of the graded Azumaya algebras over $k[T,T^{-1}]$ obtained before (see VI.5.). □

VI.6.14. Remark. The authors wonder whether there exists a short direct proof for the structure result mentioned above in Corollary VI.6.13.2., i.e. can graded-cohomology or any equivalent reasoning be avoided in deriving this result ?

To conclude let us mention :

VI.6.15. PROPOSITION. *If in the situation of VI.6.12., G is an abelian group then :*

$$H^2(G,U(S))/H^2_{gr}(G,U(S)) = H^2_{sym}(G,U(S))/H^2_{gr}(G,U(S)) \cap H^2_{sym}(G,U(S))$$

where $H^2_{sym}(G,U(S))$ *denotes the symmetric 2-cocycles* $(c(\sigma,\tau) = c(\sigma,\tau)$ *for all* $\sigma,\tau \in G)$.

PROOF. cf. [18]; similar results may be used to simplify calculation of Im $\deg^2_S$ in the case of abelian extensions S/R. □

APPENDIX. Proof of Theorem VI.6.5.

1. Exactness of the first part of the sequence :

$$1 \to U_0(R) \to U(R) \to H^0(G) \to H^1_{gr}(U) \to \mathrm{Pic}^g(R)$$

follows from Proposition VI.6.1., Theorem VI.6.2. and Proposition VI.6.4. The map $\beta : \mathrm{Pic}^g(R) \to H^0(\mathrm{Pic}^g)$ is defined by extension of scalars, i.e. $\beta([I]) = [S \underset{R}{\otimes} I]$, where $[I] \in \mathrm{Pic}^g(R)$. Exactness of the sequence at $\mathrm{Pic}^g(R)$ is then obvious.

2. Definition of $\gamma : H^0(\mathrm{Pic}^g) \to H^2_{gr}(U)$ and exactness at $H^0(\mathrm{Pic}^g)$.

Let $\underline{I} \in H^0(\mathrm{Pic}^g)$ and choose a representative $I \in [\underline{I}]$. There exists an $S^{(2)}$-isomorphism $\underline{\varphi} : I_1 \to I_2$ and by Lemma VI.2.2. this entails that there exists a $T = G(\underline{\varphi}) \in G(S^{(2)})$ and a graded $S^{(2)}$-isomorphism of degree zero : $\varphi : I_1 \to T \underset{S^{(2)}}{\otimes} I_2$. Consequently $\underline{\varphi}_2^{-1}\underline{\varphi}_3\underline{\varphi}_1$ is an $S^{(3)}$-isomorphism of I_{11}, therefore it is given by a unit $u \in U(S^{(3)})$, which is a cocycle. The graded morphism :

$$I_{11} \xrightarrow[\varphi_1]{} T_1 \underset{S^{(3)}}{\otimes} I_{13} \xrightarrow[1\otimes\varphi_3]{} T_1 \underset{S^{(3)}}{\otimes} T_3 \underset{S^{(3)}}{\otimes} I_{23} \xrightarrow[1\otimes\varphi_2^{-1}]{}$$

$$T_1 \underset{S^{(3)}}{\otimes} T_3 \underset{S^{(3)}}{\otimes} T_2^{-1} \underset{S^{(3)}}{\otimes} I_{11}$$

has degree zero. If we forget the gradations then this map is just multiplication by u, hence we obtain a graded morphism of degree zero :

$$m_u \otimes 1 : I_{11} \underset{S^{(3)}}{\otimes} I_{11}^{-1} \to T_1 \underset{S^{(3)}}{\otimes} T_2^{-1} \underset{S^{(3)}}{\otimes} T_3 \underset{S^{(3)}}{\otimes} I_{11} \underset{S^{(3)}}{\otimes} I_{11}^{-1}$$

$$m_u \quad : S^{(3)} \to \delta_1 T = T_1 \underset{S^{(3)}}{\otimes} T_2^{-1} \underset{S^{(3)}}{\otimes} T_3.$$

This entails $d_1(u) = \delta_1(T)$ and thus $[(u,T)] \in H^2_{gr}(U)$. Put $\gamma(\underline{I}) = [(u,T)]$. Let us check that γ is well-defined. If $I' \in [\underline{I}]$, $\underline{\varphi}' : I_1' \to I_2'$ and (u',T') are other choices for the objects constructed above, then there is an isomorphism $\underline{\psi} : I \to I'$ and the mapping $\underline{(\varphi')^{-1}\psi_2\, \varphi\, \psi_2^{-1}} : I_1' \to I_1'$ is an $S^{(2)}$-isomorphism and therefore it is given by multiplication by $v \in U(S^{(2)})$. Direct calculation shows that $u' = (\Delta(v))u$ and

$$d_1(u) = \underline{G((\varphi')^{-1}\psi_2\, \varphi\, \psi_1^{-1})} = (T')^{-1} \underset{S^{(2)}}{\otimes} G(\underline{\psi})_2 \underset{S^{(2)}}{\otimes} T \underset{S^{(3)}}{\otimes} G(\underline{\psi})_1^{-1}.$$

So putting $U = G(\underline{\psi})$ we obtain : $T' \underset{S^{(2)}}{\otimes} T^{-1} = d_1(v) \underset{S^{(2)}}{\otimes} \delta_0(U)$. Therefore we have $[(u,T)] = [(u',T')]$ in $H^2_{gr}(U)$.

In order to establish exactness at $H^0(\mathrm{Pic}^g)$ one first notes that $\gamma\beta = 1$. Then, if $\gamma(\underline{I}) = 1$ in $H^2_{gr}(U)$, we may write $u = \Delta_1(v)$,

$$T = d_1(v) \underset{S^{(2)}}{\otimes} \delta_0(U^{-1})$$

where (U,T) is constructed from a representative $I \in [\underline{I}]$ as above. Consider the graded morphism :

$$\varphi : I_1 \to T \underset{S^{(2)}}{\otimes} I_2 = d_1(v) \underset{S^{(2)}}{\otimes} U_1^{-1} \underset{S^{(2)}}{\otimes} U_2 \underset{S^{(2)}}{\otimes} I_2,$$

which is of degree zero.

If $m_{v^{-1}}$ is multiplication by v^{-1} then the morphism

$$m_{v^{-1}}\varphi : I_1 \underset{S^{(2)}}{\otimes} U_1 \to I_2 \underset{S^{(2)}}{\otimes} U_2,$$

is a descent datum in degree zero and it induces a graded R-module J such that $J \underset{R}{\otimes} S$ is graded isomorphic in degree zero to $I \underset{S}{\otimes} U$. But this implies that $[\underline{I}] = [\underline{J \otimes_S S}]$ in $\mathrm{Pic}^g(S)$, hence $[\underline{I}] = \beta([\underline{J}])$.

3. Definition of $\delta : H^2_{gr}(U) \to Br^g(S/R)$. Exactness at $H^2_{gr}(0)$.

Take $t = (\underline{t}, T)$, $[t] \in H^2_{gr}(U)$. Multiplication by t,

$$m_t : S^{(3)} \to d_2(t) = \delta_2(T) = T_1 \underset{S^{(3)}}{\otimes} T_2^{-1} \underset{S^{(3)}}{\otimes} T_3,$$

is a graded $S^{(3)}$-isomorphism of degree zero, and it induces a graded morphism $T_1^{-1} \underset{S^{(3)}}{\otimes} T_3^{-1} \to T_2^{-1}$ which is still given by multiplication by t. For $T \in G(S^{(2)})$ we let $P(T) = B$ be the graded S-module obtained by letting S act only on the first factor of T^{-1}. We write τ for the switch map $\tau(a \otimes b) = b \otimes a$ (earlier denoted by t but this would lead to notational inconveniences here) and consider $\tau_1 m_t$. This is not an $S^{(2)}$-isomorphism but nevertheless it induces an $S^{(2)}$-isomorphism

$$f(t) : P_1 \underset{S^{(2)}}{\otimes} T^{-1} \to P_2,$$

which is graded and of degree zero. Furthermore $f(t)$ induces a graded isomorphism of degree zero $\varphi(t) : \mathrm{END}_S(P_1) \to \mathrm{END}_S(P_2)$ as follows, $\varphi(t)(\alpha) = f(t)(\alpha \otimes 1) f(t)^{-1}$. Now we look back at the ungraded situation i.e. we consider $f = f(t) : P_1 \to P_2$. A little computation learns that $\underline{f_2^{-1} f_3 f}$ is just multiplication by some $t_{\underline{\varphi}} \in U(S^{(4)})$. Observe that $\underline{f_1} = \tau_{11} \underline{m}_{t_1}$, $\underline{f_3} = \tau_{14}\tau_{11}\underline{m}_{t_3}\tau_{11}$, $\underline{f_2} = \tau_{14}\tau_{12}\underline{m}_{t_2}$. Since $t_4 \in Z(\mathrm{End}_S\, P_{11})$ it follows that $\underline{\varphi(t)}$ is a descent datum for $\mathrm{End}_S\, \underline{P}$ and $\varphi(t)$ is a descent datum in degree zero for $\mathrm{END}_S(P)$ defining a graded Azumaya algebra

$$A(t) = \{x \in \mathrm{END}_S(P), \varphi(t)\varepsilon_1 x = \varepsilon_2 x\},$$

where the ε_i are as defined in Section VI.1., before VI.1.3. The map $s \otimes a \mapsto sa$ defines a graded isomorphism of degree zero $S \otimes A(T) \to \mathrm{END}_S(P)$, hence $[A(t)] \in Br^g(S/R)$. It is clear that $A(t)$ contains S as a maximal graded commutative subalgebra while $S \underset{R}{\otimes} S = S^{(2)}$ is a maximal graded commutative subalgebra of $\mathrm{END}_S(P)$. Put $\delta([t]) = [A(t)]$. Let us now check that δ is well-defined indeed.

a) If $t = (1, U_1 \underset{S^{(2)}}{\otimes} U_2^{-1})$ then $A(t) = \mathrm{END}_R(Q)$ for some graded R-progenerator Q !

Indeed we have $f(t) : P_1 \underset{S^{(2)}}{\otimes} U_1^{-1} \underset{S^{(2)}}{\otimes} U_2 \to P_2$. Forgetting gradations $f(t)$ becomes the identity $P_1 \to P_2$; so $f(t)$ induces a descent datum in degree zero, $P_1 \underset{S^{(2)}}{\otimes} U_1^{-1} \to P_2 \underset{S^{(2)}}{\otimes} U_2^{-1}$ which defines a graded R-progenerator Q such that the following diagram of graded $S^{(2)}$-isomorphisms of degree zero is commutative :

(*)
$$\begin{array}{ccc} Q_{13} & \to & S \underset{R}{\otimes} (P \underset{S}{\otimes} U^{-1}) \\ \downarrow & & \downarrow \\ Q_{23} & \to & (P \underset{S}{\otimes} U^{-1}) \underset{R}{\otimes} S \end{array}$$

The identity map $P \underset{S}{\otimes} U^{-1} \to P \underset{S}{\otimes} U^{-1}$ induces a graded S-isomorphism of degree zero, $\psi : END_S(P) \to END_S(P \underset{R}{\otimes} U^{-1})$ which is nothing but the identity if we forget the gradations. From (*) we may therefore derive the following commutative diagram of graded morphisms of degree zero :

$$\begin{array}{ccccc} S \underset{R}{\otimes} END_S(A) \underset{R}{\otimes} S & \to & S \underset{R}{\otimes} END_S(P \underset{S}{\otimes} U^{-1}) & \xrightarrow{1\otimes\psi^{-1}} & S \underset{R}{\otimes} END_S(P) \\ \downarrow & & \downarrow & & \downarrow \\ END_R(A) \underset{R}{\otimes} S \underset{R}{\otimes} S & \to & END_S(P \underset{S}{\otimes} U^{-1}) & \xrightarrow[\psi^{-1}\otimes 1]{} & END_S(P) \underset{R}{\otimes} S \end{array}$$

$END_R(A)$ is a descent of $END_S(P)$, hence the uniqueness in the theorem on faithfully flat descent for graded modules (Theorem VI.1.5.), entails that $END_R(Q)$ and $A(t)$ are graded isomorphic in degree zero.

b) Suppose $t = (\underline{t},T)$, $t' = (\underline{t}\Delta_1(s), T \underset{S^{(2)}}{\otimes} d_1(s))$ with $s \in U(S^{(2)})$. Then $A(t)$ and $A(t')$ are graded isomorphic in degree zero. The proof of this is formally similar to the proof in the ungraded case, so we omit it here.

c) If $t = (\underline{t},T)$, let $t^{-1} = (\underline{t}^{-1},T^{-1})$. We prove that $A(t) \underset{R}{\otimes} A(t^{-1})$ is graded isomorphic in degree zero to some $END_R(Q)$ for some graded R-progenerator Q. Consider $g = f(t) \otimes f(t^{-1})$,

$$g : P(T)_1 \underset{S^{(2)}}{\otimes} P(T^{-1}) \underset{S^{(2)}}{\otimes} T^{-1} \underset{S^{(2)}}{\otimes} T \to P(T)_2 \underset{S^{(2)}}{\otimes} P(T^{-1})_2.$$

Obviously, g induces a graded isomorphism of degree zero,

$\varphi(t) \otimes \varphi(t^{-1}) : END_S(P(T_1) \underset{S^{(2)}}{\otimes} P(T^{-1})_1) \to END_S(P(T)_2 \underset{S^{(2)}}{\otimes} P(T^{-1})_2)$,

which is a descent datum in degree zero defining $A(t) \underset{R}{\otimes} A(t^{-1})$. Note also that $P(T) \underset{S}{\otimes} P(T^{-1})$ is graded isomorphic in degree zero to $P(T_3 \underset{S^{(3)}}{\otimes} T_2^{-1})$. Using the fact that the map $f(t)_2^{-1}f(t)_3f(t)_1$ is nothing but multiplication by t_4, we easily deduce that $g_2^{-1}g_3g_1$ is nothing but multiplication by $t_{45}t_{45}^{-1}$. Hence g is a descent datum in degree zero for $P(T_3 \underset{S^{(3)}}{\otimes} T_2^{-1})$ defining a graded R-module Q making the following diagram of graded $S^{(2)}$-isomorphisms of degree zero into a commutative one :

$$\begin{array}{ccc} S \underset{R}{\otimes} Q \underset{R}{\otimes} S & \xrightarrow{1\otimes\eta} & S \underset{R}{\otimes} P(T_3 \underset{S^{(3)}}{\otimes} T_2^{-1}) \\ \downarrow & & \downarrow \\ Q \underset{R}{\otimes} S \underset{R}{\otimes} S & \xrightarrow{\eta\otimes 1} & P(T_3 \underset{S^{(3)}}{\otimes} T_2^{-1}) \underset{R}{\otimes} S \end{array}$$

This diagram leads naturally to another one :

$$\begin{array}{ccc} S \underset{R}{\otimes} END_R(Q) \underset{R}{\otimes} S & \to & END_R(P(T)_1 \underset{S^{(2)}}{\otimes} P(T^{-1})_1) \\ \downarrow & & \downarrow \\ END_R(Q) \underset{R}{\otimes} S \underset{R}{\otimes} S & \to & END_R(P(T)_2 \underset{S^{(2)}}{\otimes} P(T^{-1})_2). \end{array}$$

Comparing the latter to the commutative diagram defining $A(t) \underset{R}{\otimes} A(t^{-1})$ it is clear that this Azumaya algebra will be graded isomorphic in degree zero to $END_R(Q)$.

d) Consider s,t such that $[s],[t] \in H^2_{gr}(U)$. The argumentation expounded in c) leads to $A(s) \underset{R}{\otimes} A(t) \underset{R}{\otimes} A((st)^{-1}) \cong END_R(Q)$, the isomorphism being a graded isomorphism of degree zero. This finishes the proof of the fact that δ is well-defined.

The reader may verify directly that $\delta\gamma = 1$. So to conclude exactness at $H^2_{gr}(U)$ we consider $t \in H^2_{gr}(U)$ such that $\delta(t) = 1$ and have to show that $[(\underline{t},T)] = \gamma([\underline{J}])$ for some $[\underline{J}] \in H^0(Pic^g)$. Let $f : A(t) \to END_R(Q)$ be a graded isomorphism of degree zero, for some graded R-progenerator Q. Consider $\eta : S \underset{R}{\otimes} A(t) \to END_S(P)$ (where $P(T) = P$, $t = (\underline{t},T)$) and consider also $\psi = (1\otimes f)\eta^{-1} : END_S(P(T)) \to END_S(S \underset{R}{\otimes} Q)$. By Lemma VI.2.2. this ψ

is induced by $h : P \underset{S}{\otimes} I \to S \underset{R}{\otimes} Q$, where I is a graded S-progenerator of rank one. Consider the graded morphism of degree zero, g say, which is defined by the commutativity of the following diagram :

$$\begin{array}{ccc} S \underset{R}{\otimes} S \underset{R}{\otimes} Q & \xrightarrow[h_1]{} & P_1 \underset{S^{(2)}}{\otimes} I_1 = S \underset{R}{\otimes} (P \underset{S}{\otimes} I) \\ \downarrow t_1 & & \downarrow \\ S \underset{R}{\otimes} Q \underset{R}{\otimes} S & \xrightarrow[h_2]{} & P_2 \underset{S^{(2)}}{\otimes} I_2 = (P \underset{S}{\otimes} I) \underset{R}{\otimes} S \end{array}$$

Clearly g is a descent datum in degree zero for $P \underset{S}{\otimes} I$ defining an R-progenerator Q. One verifies :

$$\psi_3^{-1} t_1 \psi_1 = \eta_3 t_{13} t_1 t_{11}^{-1} \eta_1 = \eta_3 t_1 \eta_1^{-1} = \varphi(t).$$

Hence $g : (S \underset{R}{\otimes} P) \underset{S^{(2)}}{\otimes} I_1 \to (P \underset{R}{\otimes} S) \underset{S^{(2)}}{\otimes} I_2$, as well as

$$f(t) : (S \underset{R}{\otimes} P) \underset{S^{(2)}}{\otimes} T^{-1} \to P \underset{R}{\otimes} S,$$

induce the same map $\varphi(t)$. By Lemma IV.2.2., there is a graded $S^{(2)}$-isomorphism of degree zero $\varphi : I_1 \underset{S^{(2)}}{\otimes} I_2^{-1} \to T^{-1}$, such that $g = f(t) \underset{2}{\otimes} \varphi$. If one writes down the descent condition for g one obtains that $t_1 t_2^{-1} t_3 \varphi_2^{-1} \varphi_3 \varphi_1$ is the identity map of $(P \underset{S}{\otimes} I)_{11}$. Since $t_1 t_2^{-1} t_3$ is just multiplication by t_4, whilst $\varphi_2^{-1}\varphi_3\varphi_1$ is just multiplication by t, it follows that $\gamma(\underline{I}^{-1}) = [(\underline{t},T)]$.

4. <u>Definition of</u> $\zeta : Br^g(S/R) \to H^1(Pic^g)$. <u>Exactness at</u> $Br^g(S/R)$.

Let $[A] \in Br^g(S/R)$ and let $\tau : S \underset{R}{\otimes} A \to END_S(Q)$ be a gr-splitting for A. Denote by φ the morphism $\tau_3 \tau_2 \tau_1^{-1} : END_{S^{(2)}}(Q_1) \to END_{S^{(2)}}(Q_2)$. This φ is graded of degree zero and it is induced by a graded isomorphism of degree zero $f : Q_1 \underset{S^{(2)}}{\otimes} I \to Q_2$ where I is a graded $S^{(2)}$-progenerator of rank one. Since $\varphi_2 = \varphi_3\varphi_1$ we have $I_2 = I_3 \underset{S^{(3)}}{\otimes} I_1$ in $Pic^g(S^{(2)})$ and so we may write $\zeta([A]) = [\underline{I}] \in H^1(Pic^g)$. It is straightforward to verify that ζ is well-defined and that $\zeta\delta = 1$. Next, suppose that $\zeta([A]) = 1$ in $H^1(Pic^g)$ i.e. $[\underline{I}] = 1$ where $\underline{I}$ is constructed from A as indicated above. There is an $S^{(2)}$-isomorphism $\psi : \underline{J_1} \underset{S^{(2)}}{\otimes} \underline{J_2^{-1}} \to \underline{I}$, where J is a graded S-progenerator of rank one. Consider

$$\underline{\psi} \otimes 1 : \underline{I} \underset{S^{(2)}}{\otimes} \underline{J_1^{-1}} \underset{S^{(2)}}{\otimes} \underline{J_2} \to S^{(2)},$$

and put $T = G(\underline{\psi} \otimes 1)$. Then we obtain a graded $S^{(2)}$-module isomorphism of degree zero, $\psi : I \to J_1 \underset{S^{(2)}}{\otimes} J_2^{-1} \underset{S^{(2)}}{\otimes} T$. Again by Lemma VI.2.2., we may generate φ by a graded $S^{(2)}$-isomorphism of degree zero

$$g : (Q \underset{S}{\otimes} J)_1 \underset{S^{(2)}}{\otimes} T \to (Q \underset{S}{\otimes} J)_2.$$

Put $U = Q \underset{S}{\otimes} J$. Consider

$$g_2^{-1} g_3 g_1 : U_{11} \underset{S^{(3)}}{\otimes} T_1 \underset{S^{(3)}}{\otimes} T_3 \underset{S^{(3)}}{\otimes} T_2^{-1} \to$$

$$\to U_{21} \underset{S^{(3)}}{\otimes} T_3 \underset{S^{(3)}}{\otimes} T_2^{-1} = U_{13} \underset{S^{(3)}}{\otimes} T_3 \underset{S^{(3)}}{\otimes} T_2^{-1} \to U_{23} \underset{S^{(3)}}{\otimes} T_2^{-1} \to U_{11}.$$

The ungraded $\underline{g_2^{-1} g_3 g_1}$ is an $S^{(3)}$-automorphism of U_{11}, hence it is given by multiplication by $\underline{t} \in U(S^{(3)})$ and $\underline{t}$ is easily seen to be a cocycle. However the following maps are graded $S^{(3)}$-isomorphisms of degree zero :

$$g_2^{-1} g_3 g_1 : U_{11} \underset{S^{(3)}}{\otimes} \delta_1(T) \to U_{11}$$

$$g_2^{-1} g_3 g_1 : U_{11} \underset{S^{(3)}}{\otimes} d_2(t) \to U_{11}.$$

The fact that U_{11} is a graded faithfully flat $S^{(3)}$-module entails that $d_2(t)$ and $\delta_1(T)$ are graded isomorphic in degree zero, hence $t = (\underline{t}, T) \in H^2_{gr}(U)$. Let as in 3., $\eta(t^{-1})$ be a canonical graded splitting of $A(t^{-1})$ and write P for $P(T^{-1})$. Consider,

$$\tau \underset{S}{\otimes} \eta(t^{-1}) : S \underset{R}{\otimes} A \underset{R}{\otimes} A(t^{-1}) \to END_S(Q \underset{S}{\otimes} P).$$

Then $\varphi \otimes \varphi(t^{-1})$ is induced by :

$$h = g \otimes f(t^{-1}) : (U \underset{S}{\otimes} P)_1 \to (U \underset{S}{\otimes} P)_2.$$

But $h_2^{-1} h_3 h_1$ is just multiplication by $tt^{-1} = 1$, so h is a descent datum in degree zero for $U \underset{S}{\otimes} P$, defining the descended R-module M say.

The uniqueness in the theorem of faithfully flat descent for graded modules entails that $A \otimes_R A(t^{-1})$ and $END_R(M)$ are graded isomorphic in degree zero. Consequently $[A] = [A(t^{-1})]$ in $Br^g(S/R)$.

5. <u>Definition of $\eta : H^1(Pic^g) \to H^3_{gr}(U)$.</u> <u>Exactness at $H^1(Pic^g)$.</u>

If $[\underline{I}] \in H^1(Pic^g)$, choose $I \in [\underline{I}]$ together with an $S^{(3)}$-isomorphism $\underline{f} : I_1 \otimes_{S^{(3)}} I_3 \to I_2$. Put $T = G(\underline{f}) \in G(S^{(3)})$ and then we obtain a graded $S^{(3)}$-morphism of degree zero :

$$f : I_1 \otimes_{S^{(3)}} I_3 \to T \otimes_{S^{(3)}} I_2.$$

We look at :

$$f_4^{-1}f_2^{-1}f_3f_1 : I_{11} \otimes_{S^{(4)}} I_{31} \otimes_{S^{(4)}} I_{33} \to I_{11} \otimes_{S^{(4)}} I_{31} \otimes_{S^{(4)}} I_{33} \otimes_{S^{(4)}} \delta_2(T).$$

Exactly as in 4., we see that this map is just multiplication by $\underline{t} \in U(S^{(4)})$, $d_3(\underline{t}) = \delta_3(T)$ and $\underline{t}$ is a cocycle. So we define $\eta([\underline{I}]) = t = (\underline{t},T)$. Verification of the fact that η is well-defined as well as $\eta\zeta = 1$ is straightforward.

Now suppose $\eta([\underline{I}]) = [t] = 1$ in $H^3_{gr}(U)$. So there exists $\underline{v} \in U(S^{(3)})$ and $V \in G(S^{(2)})$ such that $\underline{t} = \Delta_2\underline{v}$, $T = d_2(\underline{v}) \otimes_{S^{(3)}} \delta_1(V^{-1})$. This yields a graded $S^{(3)}$-isomorphism of degree zero :

$$f : I_1 \otimes_{S^{(3)}} I_3 \to I_2 \otimes_{S^{(3)}} d_2(\underline{v}) \otimes_{S^{(3)}} \delta_1(V^{-1}).$$

Composing f and $m_{V^{-1}}$ we get $f' = fm_{V^{-1}} : I_1' \otimes_{S^{(3)}} I_3' \to I_2'$ where $I' = I \otimes_{S^{(2)}} U$. Clearly $(f_2')^{-1}f_3'f_1' = f_4'$.

We consider Q and I' as graded S-modules with S acting only on the first factor (i.e. : $s.a = (s \otimes 1)a$). The composition of morphisms of abelian groups,

$$g : Q_1 \otimes_{S^{(2)}} I' = (S \otimes_R Q) \otimes_{S^{(3)}} I_3' \xrightarrow{f'} I_2' \xrightarrow{t_1} Q_2$$

is then a graded $S^{(2)}$-isomorphism of degree zero and $g_2^{-1}g_3g_1 = (f_2')^{-1}f_3'f_1'$ because : $g_1 = t_{11}f_1'$, $g_3 = t_{14}f_4' = t_{14}t_{11}f_3't_{11}$, $g_2 = t_{14}t_{12}f_2'$ (here t_{ij} denoting the switch maps).

However, $f_4' : Q_{11} \underset{S(4)}{\otimes} I'_{14} \underset{S(4)}{\otimes} I'_{34} \to Q_{11} \underset{S(4)}{\otimes} I'_{24}$ induces the identity of $END_{S^{(3)}}(Q_{11})$. This means that conjugation by g determines a descent datum in degree zero defining an Azumaya algebra A over R. Obviously $\zeta([A]) = [\underline{I}'] = [\underline{I}]$. □

<u>Remark</u>. We have reproduced S. Caenepeel's proof here, cf. [16]. Of course this proof is, up to the graded peculiarities, rather close to the original proof as it is given e.g. in [56].

CHAPTER VII
APPLICATIONS IN ALGEBRAIC GEOMETRY

VII.1. The Brauer Group of a Projective Variety.

In [47] A. Grothendieck introduced and studied the Brauer group of a scheme. Similar constructions for arbitrary ringed spaces were investigated by B. Auslander in [8]. The necessary background on these sheaf theoretical methods has been provided in Section IV.1. In Section IV.2., Theorem IV.2.8. we showed that the Brauer group of a quasi-affine open subscheme X(I) of X = Spec(R) is nothing but the relative Brauer group $Br(R,\kappa_I)$. The aim of this Chapter VII is to obtain a ring theoretical description of the Brauer group of a projective scheme or variety. Actually we will consider the ringed space Proj(R) where R is a positively graded ring (see also VI.1. after Lemma VI.1.1.). The assumption that Br(Proj(R)) reduces to the graded Brauer group $Br^g(R)$ of R is incorrect in general; actually it is this fact that prompted the introduction of relative (graded) Brauer groups.

Throughout $R = \bigoplus_{n=0}^{\infty} R_n$ is a commutative positively graded ring with unit, R_+ is the ideal $\bigoplus_{n=1}^{\infty} R_n$ and Proj(R) = {P graded prime ideal of R; $P \not\supset R_+$}. To a graded ideal I of R we associate the set $V_+(I) = \{P \in Proj(R);\ I \not\subset P\}$. The Zariski topology on Proj(R) is given by taking the sets $X_+(I) = Proj(R) - V_+(I)$ for the open sets. If P is a graded prime ideal of R then we have defined the graded ring of fractions $Q^g_{R-P}(R)$ in Section VI.1. To any graded ideal I of R we associate a

graded kernel functor κ_I^g with idempotent filter $C^g(I) = \{J$ ideal of R; $J \supset I^m$ for some m$\}$.
The graded ring of quotients $Q_I^g(R)$ may be defined as $\varinjlim_m \mathrm{HOM}_R(I^m, R)$ or as the subring of $Q_I(R)$ defined by $Q_I^g(R)_n = \{x \in Q_I(R);\ J_p x \subset R_{p+n}$ for some graded ideal J of R, and for all $p \in \mathbb{N}\}$ and $Q_I^g(R) = \bigoplus_n Q_I^g(R)_n$. We refer to [65] or [87] for full detail (which is not needed here) on graded localization.
We may construct a sheaf of graded algebras $\mathcal{O}_R^+$ on Proj(R) by putting $\Gamma(X_+(I), \mathcal{O}_R^+) = Q_I^g(R)$ and using the obvious localization morphisms as restriction maps (these are graded morphisms of degree zero). The stalk $\mathcal{O}_{R,P}^+$ of $\mathcal{O}_R^+$ at $P \in \mathrm{Proj}(R)$ is exactly $Q_{R-P}^g(R)$ as an easy limit calculation shows. We say that $\mathcal{O}_R^+$ is the <u>graded structure sheaf</u> on Proj(R). The <u>structure sheaf</u> of Proj(R) is defined to be $\tilde{R} = (\mathcal{O}_R^+)_0$ i.e. $\Gamma(X_+(I), \tilde{R}) = (Q_I^g(R))_0$. For an arbitrary graded R-module M we construct $Q_I^g(M)$ resp. $Q_{R-P}^g(M)$ as well as $\mathcal{O}_M^+$ and $\tilde{M}$ in a similar way. For notational convenience we will write $M_{(P)}$ for $(Q_{R-P}^g(M))_0$.

Recall that for any $n \in \mathbb{Z}$, the shift functor T_n is defined by $(T_n M)_p = M_{n+p}$ for all $p \in \mathbb{Z}$, $M \in R\text{-gr}$. The sheaf of graded $\mathcal{O}_R^+$-modules, resp. $\tilde{R}$-modules, associated with $T_n M$ will be denoted by $\mathcal{O}_R^+(n)$, resp. $\tilde{R}(n)$. For any sheaf of $\tilde{R}$-modules $\mathcal{M}$ we put $\mathcal{M}(n) = \mathcal{M} \otimes_{\tilde{R}} \tilde{R}(n)$. We define a left exact functor Γ_* from sheaves of $\tilde{R}$-modules to graded R-modules, by putting $\Gamma_* \mathcal{M} = \bigoplus_{n \in \mathbb{Z}} \Gamma(\mathrm{Proj}(R), \mathcal{M}(n))$. We assume from hereon that R is generated as an R_0-algebra by a finite subset of R_1. Let us recall some well-known facts, cf. for example [96].

VII.1.1. <u>PROPOSITION</u>. 1. *For any quasicoherent sheaf of* $\tilde{R}$-*modules the canonical morphism* $\beta : \Gamma_*(\mathcal{M})^\sim \to \mathcal{M}$ *is an isomorphism.*

2. *Every quasicoherent sheaf of* $\tilde{R}$-*modules (of finite type)* $\mathcal{M}$ *is of the form* $\tilde{M}$ *for some (finitely generated) graded* R-*module* M.

3. *If* $M, N \in R\text{-gr}$ *are such that* M *is finitely generated then there is a canonical isomorphism* $\mu : (\mathrm{HOM}_R(M,N))^\sim \to \mathcal{H}om_{\tilde{R}}(\tilde{M}, \tilde{N})$.

4. *For any* $M, N \in R\text{-gr}$ *there is a canonical isomorphism* $\lambda : \tilde{M} \otimes_{\tilde{R}} \tilde{N} \to (M \otimes_R N)^\sim$.

5. *If* $M \in R\text{-gr}$ *then* $\widetilde{M} = 0$ *if and only if for all* $m \in M$ *and* $r \in R_n$, $n \geq 1$, *we may find a* $p \in \mathbb{N}$ *such that* $r^p m = 0$.

6. *For any* $M \in R\text{-gr}$, $(T_n M)^{\sim} = \widetilde{M}(n)$.

7. *Suppose that* R_0 *is a finitely generated algebra over a field and let* M *be a finitely generated graded* R*-module. The canonical map* $\alpha : M \to T(\widetilde{M})$ *induces isomorphisms* $\alpha_\alpha : M_\alpha \to \Gamma(\text{Proj}(R), \widetilde{M}(d))$ *for* d *large enough.* □

VII.1.2. PROPOSITION. *Consider* $M,N \in R\text{-gr}$ *and let* M *be finitely presented. Write* E *for* $HOM_R(M,N) = Hom_R(M,N)$. *There is a canonical isomorphism :*

$$\mu : \mathcal{O}_E^+ \to \mathcal{HOM}_{\mathcal{O}_R^+}(\mathcal{O}_M^+, \mathcal{O}_N^+) = \mathcal{B}.$$

PROOF. Recall that $\mathcal{B}$ is the sheaf defined by putting

$$\Gamma(U,\mathcal{B}) = HOM_{\mathcal{O}_R^+|U}(\mathcal{O}_M^+|U, \mathcal{O}_N^+|U),$$

where for arbitrary sheaves of graded modules M and N over some sheaf of graded rings R we define $HOM_R(M,N)$ to consists of all "compatible" families $\{f_U;\ U \text{ open in } X,\ f_U \in Hom_{R(U)}(M(U),N(U))\}$. Now, if $f \in R_d$ for some $d \in \mathbb{N}$ then we can define a morphism $\mu_{f,p}$(where Q_f^g is the graded localization functor obtained by inverting the homogeneous elements $\{1,f,f^2,\ldots\}$) by

$$\mu_{f,p} : Q_f^g(HOM_R(M,N))_p \to HOM_{Q_f^g(R)}(Q_f^g(M), Q_f^g(N))_p$$

by taking for $\mu_{f,p}(uf^{-n})$ the morphism sending xf^{-r} to $u(x)f^{-r-n}$, where $u \in HOM_R(M,N)_{nd+p}$. Let μ_f be $\bigoplus_p \mu_{f,p}$. If $h \in R_e$ then we obtain a commutative diagram of graded morphisms of degree zero :

$$\begin{array}{ccc} Q_f^g(HOM_R(M,N)) & \xrightarrow{\mu_f} & HOM_{Q_f^g(R)}(Q_f^g(M), Q_f^g(N)) \\ \downarrow & & \downarrow \\ Q_{hf}^g(HOM_R(M,N)) & \xrightarrow{\mu_{hf}} & HOM_{Q_{hf}^g(R)}(Q_{hf}^g(M), Q_{hf}^g(N)) \end{array}$$

This permits to glue the morphisms μ_f together over Proj(R) and so we obtain the desired morphism $\mu : \mathcal{O}_E^+ \to \mathcal{B}$. To check that μ is an isomorphism is fairly easy and we leave this to the reader. □

VII.1.3. Remarks.

1. On the structure sheaf level the above yields that for finitely presented $M \in R$-gr we have $\mathcal{H}om_{\tilde{R}}(\tilde{M},\tilde{N}) = (\mathcal{HOM}_{\mathcal{O}_R^+}(\mathcal{O}_M^+,\mathcal{O}_N^+))_0$.

2. Since we assumed that R_1 generates R as an R_0-algebra we have that $Q_P^g(R) \cong R_{(P)}[T,T^{-1}]$ where T is an indeterminate of degree one. We may use all results concerning strongly graded rings for the stalks of the graded structure sheaf (see Section VI.1.), in particular a graded projective $Q_P^g(R)$-module M is of the form $M_0 \otimes_{R_{(P)}} Q_P^g(R)$ for some projective $R_{(P)}$-module M_0.

The terminology we use is that of Section IV.1. In particular we recall Proposition IV.1.1. and Lemma IV.1.2. (which was noted to hold for arbitrary schemes); in a similar vein we now have, with notations as before,

VII.1.4. LEMMA. *If $\mathcal{M}$ is a locally projective sheaf of $\tilde{R}$-modules of finite type, then there exists a graded R-module M of finite presentation such that $\mathcal{M} = \tilde{M}$.*

PROOF. Since we already have seen that $\mathcal{M} = \tilde{N}$ for some finitely generated graded R-module N, the lemma is trivial in case R is Noetherian. If R is not Noetherian the lemma still holds, we refer to [42] for a proof. □

VII.1.5. PROPOSITION. *Let $\mathcal{M},\mathcal{N}$ be locally projective sheaves of finite type over $\tilde{R}$ and put $\mathcal{M}^+ = \mathcal{M} \otimes_{\tilde{R}} \mathcal{O}_R^+$, $\mathcal{N}^+ = \mathcal{N} \otimes_{\tilde{R}} \mathcal{O}_R^+$, then we have :*

$$\mathcal{H}om_{\tilde{R}}(\mathcal{M},\mathcal{N}) \otimes_{\tilde{R}} \mathcal{O}_R^+ = \mathcal{HOM}_{\mathcal{O}_R^+}(\mathcal{M}^+,\mathcal{N}^+).$$

PROOF. Since there is a canonical sheaf morphism

$$\lambda : Hom_{\widetilde{R}}(M,N) \underset{\widetilde{R}}{\otimes} O_R^+ \to HOM_{O_R^+}(M^+,N^+)$$

it will be sufficient to check that for each $P \in \mathrm{Proj}(R)$ the local morphism λ_P at P is an isomorphism. By the lemma we may take $M = \widetilde{M}$, $N = \widetilde{N}$ for M and N $\in$ R-gr which are also finitely presented R-modules. On one hand we have :

$$(Hom_{\widetilde{R}}(\widetilde{M},\widetilde{N}) \underset{\widetilde{R}}{\otimes} O_R^+)_P = (Hom_{\widetilde{R}}(\widetilde{M},\widetilde{N})_P \underset{\widetilde{R}_P}{\otimes} O_{R,P}^+ =$$

$$= (\mathrm{HOM}_R(M,N)^{\sim})_P \underset{R_{(P)}}{\otimes} Q_P^g(R) = \mathrm{HOM}_R(M,N)_{(P)} \underset{R_{(P)}}{\otimes} Q_P^g(R).$$

On the other hand, since M is finitely presented :

$$(Hom_{\widetilde{R}}(M,N) \underset{\widetilde{R}}{\otimes} O_R^+)_P = \mathrm{Hom}_{\widetilde{R}_P}(M_P,N_P) \underset{R_{(P)}}{\otimes} Q_P^g(R)$$

$$= \mathrm{Hom}_{R_{(P)}}(M_{(P)},N_{(P)}) \underset{R_{(P)}}{\otimes} Q_P^g(R)$$

$$= \mathrm{Hom}_{Q_P^g(R)}(M_{(P)} \underset{R_{(P)}}{\otimes} Q_P^g(R), N_{(P)} \underset{R_{(P)}}{\otimes} Q_P^g(R))$$

$$= \mathrm{Hom}_{Q_P^g(R)}(Q_P^g(M),Q_P^g(N)) = \mathrm{Hom}_{Q_P^g(R)}(Q_P^g(M),Q_P^g(N))$$

(using the fact that $Q_P^g(M)$ is finitely presented over $Q_P^g(R)$).
A similar argument yields $Q_P^g(M) = (\widetilde{M} \underset{\widetilde{R}}{\otimes} O_R^+)_P$ and $Q_P^g(N) = (\widetilde{N} \underset{\widetilde{R}}{\otimes} O_R^+)_P$, so finally :

$$(Hom_{\widetilde{R}}(M,N) \underset{\widetilde{R}}{\otimes} O_R^+) = \mathrm{HOM}_{Q_P^g(R)}((M^+)_P,(N^+)_P) = (HOM_{O_R^+}(M^+,N^+))_P.$$

□

Let $\kappa_{R_+}^g = \kappa$ be the graded kernel functor associated to the filter L_+ generated by the powers of R_+ and let $Q_{R_+}^g(-)$ denote the associated graded localization functor introduced before (taking $I = R_+$). Since R_+ is generated by a finite number of elements of degree one it follows, even in the non-Noetherian case, that κ is of finite type (in the sense of [43]; [44]) and that $Q_{R_+}^g(-)$ commutes with direct sums (cf. loc. cit.).

VII.1.6. PROPOSITION. *If M is a locally projective* $\widetilde{R}$*-module of finite type then* $\Gamma_*(M)$ *is* κ*-finitely presented.*

PROOF. By Lemma VII.1.4., $\mathcal{M} = \tilde{M}$ for some graded finitely presented R-module M. Choose a finite presentation for M, say $F_1 \xrightarrow{\psi} F_1 \to M \to 0$, where F_1, F_2 are gr-free R-modules of finite rank. The properties of $Q^g_{R_+}(-)$ mentioned in the remarks preceding this proposition, entail that Coker $Q^g_{R_+}(\psi)$ is finitely presented in $Q^g_{R_+}(R)$-gr. The fact that $\mathcal{M} = (\Gamma_*(\mathcal{M}))^\sim$ combined with the property $\kappa = \inf\{\kappa^g_{R-P}: P \in \mathrm{Proj}(R)\}$ yields the result. □

The foregoing proposition brings relativety back into the picture. Before making use of this fact let us describe the relation between $Q^g_{R_+}(-)$ and Γ_*.

VII.1.7. PROPOSITION. *If* $M \in$ R-gr *then* $Q^g_{R_+}(M) = \Gamma_*(\tilde{M})$.

PROOF. Consider the following commutative diagram of graded R-module morphisms of degree zero :

$$\begin{array}{ccc} M & \xrightarrow{j} & Q^g_{R_+}(M) \\ \alpha \downarrow & & \downarrow \beta \\ \Gamma_*(\tilde{M}) & \xrightarrow[\Gamma_*(\tilde{j})]{} & \Gamma_*((Q^g_{R_+}(M))^\sim) \end{array}$$

It is clear that $\Gamma_*(\tilde{j})$ is completely determined by j, i.e. $\Gamma_*(\tilde{j}) = \varphi$ is the unique graded morphism making the diagram commute. Write $N = Q^g_{R_+}(M)$. Note that $\mathcal{O}^+_M = \mathcal{O}^+_N$. Indeed the localization morphism j induces a morphism $\mathcal{O}^+_j : \mathcal{O}^+_M \to \mathcal{O}^+_N$ in the usual way. For any $P \in \mathrm{Proj}(R)$, the local morphism $\mathcal{O}^+_{j,P} : \mathcal{O}^+_{M,P} \to \mathcal{O}^+_{N,P}$ reduces to $Q^g_{R-P}(j) : Q^g_{R-P}(M) \to Q^g_{R-P}(N)$. Since $R_+ \not\subset P$ it follows that $\kappa \leq \kappa^g_{R-P}$, hence $Q^g_{R-P}(N) = Q^g_{R-P}(M)$ and $Q^g_{R-P}(j)$ is the identity. Consequently $\mathcal{O}^+_j$ is the identity too. Then $\mathcal{O}^+_M = \mathcal{O}^+_N$ entails $\tilde{M} = \tilde{N}$ and so $\Gamma_*(\tilde{j})$ is actually the identity on $\Gamma_*(\tilde{M})$. For any $P \in$ R-gr the global sections of $\mathcal{O}^+_P$ are given by

$$\Gamma(\mathrm{Proj}(R), \mathcal{O}^+_P) = \varinjlim_n \mathrm{HOM}_R(R^n_+, P) = Q^g_{R_+}(P).$$

It is equally obvious that $\mathcal{O}_M^+ = \mathcal{O}^+_{\Gamma_*(\widetilde{M})}$ and therefore we obtain a canonical morphism :

$$\gamma : \Gamma_*(\widetilde{M}) \to \Gamma(\mathrm{Proj}(R), \mathcal{O}^+_{\Gamma_*(\widetilde{M})}) = \Gamma(\mathrm{Proj}(R), \mathcal{O}^+_M) = Q^g_{R_+}(M).$$

Let us look at the following diagrams in R-gr :

Now : $(\beta\gamma)\alpha = \beta(\gamma\alpha) = \beta j = \alpha$, hence $\beta\gamma = 1_{\Gamma_*(\widetilde{M})}$ (because the extension $\Gamma_*(\tilde{j})$ of j is unique). On the other hand, $(\gamma\beta)j = \gamma\alpha = j$, hence $\gamma\beta = 1_N$ (because the identity of N is the unique morphism extending the identity of $j(M)$). It follows that $\Gamma_*(\widetilde{M}) = Q^g_{R_+}(M)$. □

VII.1.8. <u>COROLLARY</u>. *Let* A *be any commutative ring and*

$$X = \mathbb{P}^n_A = \mathrm{Proj}(A[X_0,\ldots,X_n]),$$

then $\Gamma_*(\mathcal{O}_X) = A[X_0,\ldots,X_n]$.

<u>PROOF</u>. It is easy to see that $Q^g_{R_+}(R) = R$ if $R = A[X_0,\ldots,X_n]$ and $R_+ = (X_0,\ldots,X_n)$. □

The foregoing proposition gives a torsion theoretic meaning to the functor Γ_*. If one is not familiar with localization gadgets like $Q^g_{R_+}(-)$ one might be happy about the foregoing proposition, hoping that it makes it possible to drop the localization techniques. However, we actually work in the opposite direction, and the above corollary already indicates in what way the localization techniques might be applied to our benefit. We first need another simple lemma.

VII.1.9. <u>LEMMA</u>. *Let* R *be any positively graded commutative ring and consider* $M,N \in$ R-gr *and a graded morphism* $u : M \to N$ *in* $\mathrm{HOM}_R(M,N)$.

1. *If* M *is finitely generated and the graded morphism*

$$Q^g_{R-P}(u) : Q^g_{R-P}(M) \to Q^g_{R-P}(N)$$

is surjective for some $P \in \mathrm{Proj}(R)$ *, then there is an* $f \in h(R)$ *such that* $P \in X_+(f)$ *and such that the graded morphism*

$$Q^g_f(u) : Q^g_f(M) \to Q^g_f(N),$$

is surjective.

2. *If* M *is finitely generated while* N *is finitely presented and*

$$Q^g_{R-P}(u) : Q^g_{R-P}(M) \to Q^g_{R-P}(N)$$

is bijective for some $P \in \mathrm{Proj}(R)$, *then there is an* $f \in h(R)$ *such that* $P \in X_+(f)$ *and such that the graded morphism*

$$Q^g_f(u) : Q^g_f(M) \to Q^g_f(N)$$

is bijective.

3. *If* N *is finitely presented such that* $Q^g_{R-P}(N)$ *is a gr-free* $Q_{R-P}(R)$*-module of finite rank, then there exists an* $f \in h(R-P)$ *such that* $Q^g_f(N)$ *is gr-free over* $Q^g_f(R)$ *of the same finite rank.*

PROOF. Easy. □

VII.1.10. PROPOSITION. *Let* R *be a positively graded commutative ring generated by* R_1 *as an* R_0*-algebra, then the following statements are equivalent :*

1. *M is locally projective of finite type in* $\tilde{R}$-Mod.
2. $\Gamma_*(M)$ *is a graded* κ*-finitely generated,* κ*-quasiprojective* $Q^g_{R_+}$*-module, where* $\kappa = \kappa^g_{R_+}$.

PROOF. 1 ⇒ 2. From Proposition VII.1.6. it follows that $\Gamma_*(M)$ is κ-finitely generated. Furthermore $(M)_P$ is a finitely generated projective $Q^g_{R-P}(R)$-module for each $P \in \mathrm{Proj}(R)$, so 2. follows.

2 ⇒ 1. Pick a graded finitely presented $M \subset \Gamma_*(M)$ such that $\Gamma_*(M)/M$ is κ-torsion. Clearly, $\tilde{M}$ is finitely presented over $(Q^g_{R_+}(R))^{\sim} = \tilde{R}$ and

$\tilde{M} = M$ since $Q^g_{R-P}(\Gamma_*(M)/M) = 0$ for all $P \in \text{Proj}(R)$. Moreover, $Q^g_{R-P}(\Gamma_*(M))$ is a projective $Q^g_{R-P}(R)$-module, hence $M_{(P)} = (Q^g_{R-P}(\Gamma_*(M)))_0 = \Gamma_*(M)_{(P)}$ is a projective $\tilde{R}_P = R_{(P)}$-module by the earlier results. Thus, M is a locally projective Module of finite type. □

At this point let us state the graded version of Lemma III.3.1., the proof is an easy modification of the proof given in Section III.3.

VII.1.11. LEMMA. *Let* A *be a graded* R*-algebra. Consider a family of graded kernel functors* $\{\kappa_\alpha;\ \alpha \in I\}$ *such that each* $Q^g_{\kappa_\alpha}(A)$ *has center* $Q^g_{\kappa_\alpha}(R)$. *Write* $\kappa = \inf\{\kappa_\alpha;\ \alpha \in I\}$; *then* $Q^g_\kappa(A)$ *has center* $Q^g_\kappa(R)$. □

VII.1.12. LEMMA. *Let* S *be a commutative graded* R*-algebra which is a strongly graded ring i.e.* $SS_1 = S$. *If* A *is a graded* S*-algebra such that* $Z(A_0) = S_0$, *then* $Z(A) = S$.

PROOF. Since Z(A) is graded we only have to consider an homogeneous $x \in Z(A)_d$ and show that $x \in S_d$. Clearly $S_{-d}x \subset Z(A)_0 \subset Z(A_0) = S_0$ hence $SS_{-d}x = Sx \in S$ or $x \in S$. □

VII.1.13. COROLLARY.

1. *If for some graded* R*-algebra* A *we know that* $\tilde{A}$ *is a central sheaf of* $\tilde{R}$*-algebras then* O^+_A *is a central sheaf of* O^+_R*-algebras.*
2. *If for some graded* R*-algebra* A *we have that* A *is a locally separable sheaf of* $\tilde{R}$*-algebras then for each* $P \in \text{Proj}(R)$, $Q^g_{R-P}(A)$ *is central separable over* $Q^g_{R-P}(R)$.

PROOF. 1. Is obvious. 2. Follows directly from the fact that

$Q^g_{R-P}(R) = R_{(P)}[T,T^{-1}]$ and $Q^g_{R-P}(A) = A_{(P)} \otimes_{R_{(P)}} Q^g_{R-P}(R)$.

In the sequel of this section R is positively graded commutative, Noetherian, generated as an R_0-algebra by (a finite number of elements from) R_1.

By the above remarks and lemmas it follows that each finitely presented central sheaf A of $\tilde{R}$-algebras over Proj(R) yields a graded central $\Gamma_*(\tilde{R})$-algebra. Since we assumed R (so also R_0) to be Noetherian, any sheaf of $\tilde{R}$-modules M of finite type is of the form $\tilde{M}$ for some $M \in R\text{-gr}$ which is finitely presented and κ-torsion free ($\kappa = \kappa^g_{R_+}$). Indeed $M = \tilde{N}$ for some finitely presented N, then put $M = N/\kappa(N)$ and check $M = \tilde{M}$.

VII.1.14. LEMMA. *Let* L *be a graded* κ*-quasiprojective* κ*-torsion free* R*-module of finite presentation and put* $L = \tilde{L}$*, then :*

$$\mathrm{Hom}_{\Gamma_*(\tilde{R})}(\Gamma_*(L),\Gamma_*(L)) = \Gamma_*(\mathcal{H}om_{\tilde{R}}(L,L)).$$

PROOF. Since $\Gamma_*(L) = Q^g_{R_+}(L)$ and $\Gamma_*(\tilde{R}) = Q^g_{R_+}(R)$ by Proposition VII.1.7. this is just a straightforward graded version of Lemma III.1.5. (taking into account Proposition VII.1.1.,3). We leave details to the reader. □

VII.1.15. COROLLARY. *Let* L *be a sheaf of* $\tilde{R}$*-modules which is locally of finite type, then*

$$\mathrm{Hom}_{\Gamma_*(\tilde{R})}(\Gamma_*(L),\Gamma_*(L)) = \Gamma_*(\mathcal{H}om_{\tilde{R}}(L,L)). \qquad \square$$

VII.1.16. LEMMA. *If* L *and* L' *are sheaves of* $\tilde{R}$*-modules then :*

$$\Gamma_*(L \underset{\tilde{R}}{\otimes} L') = Q^g_{R_+}(\Gamma_*(L) \underset{\Gamma_*(\tilde{R})}{\otimes} \Gamma_*(L')) = \Gamma_*((\Gamma_*(L) \underset{\Gamma_*(\tilde{R})}{\otimes} \Gamma_*(L'))^{\sim}).$$

□

Now we extend some of the terminology already introduced in the ungraded or the "reflexive" graded case as follows.

A graded R-algebra A is said to be a κ-Azumaya algebra if it is κ-closed, κ-quasiprojective and κ-finitely presented, with $\kappa = \kappa^g_{R_+}$, and if the canonical graded morphism of degree zero $A^e = A \underset{R}{\otimes} A^0 \to \mathrm{HOM}_R(A,A)$ induces an isomorphism of degree zero : $Q^g_{R_+}(A^e) \cong \mathrm{HOM}_R(A,A)$. We have established :

VII.1.17. THEOREM. *There is a bijective correspondence between locally separable central sheaves of* $\tilde{R}$*-algebras over* Proj(R) *and graded* $\kappa^g_{R_+}$*-Azumaya algebras over* R. □

It is clear how $Br^g(R,\kappa^g_{R_+})$ is defined (one runs along the lines of the definition of $Br(R,\kappa)$ or $\beta^g(R)$ given in ealier Chapters). So finally the above theorem may be translated to :

VII.1.18. COROLLARY. *Let* R *be a commutative Noetherian positively graded ring generated by* R_1 *as an* R_0*-algebra. Then* $Br(Proj(R)) = Br^g(R,\kappa^g_{R_+})$. □

Of course this equality can be further exploited in studying Brauer groups of projective varieties but this would carry us to far (away from the more ring theoretical purposes of this book).
We will however treat in some more detail the case of projective curves.

VII.2. Brauer Groups of Projective Curves.

In this section we extend further in the one-dimensional case the ideas of Section VII.1.

Recall that a scheme X is said to be normal if its local rings are integrally closed domains. If X is an affine scheme then it is normal exactly then when its affine coordinate ring is integrally closed. If X is one-dimensional then X is normal exactly then when it is a nonsingular variety.

Let C be a commutative ring and consider a closed subscheme of $\mathbb{P}^r_C$ (see R. Hartshorne [49] and some definitions in the foregoing section). Let $\mathcal{I}_X$ be the sheaf of ideals defining X, write $I = \Gamma_*(\mathcal{I}_X)$, and let S[X] be the ring $C[X_0,\ldots,X_r]/I$. We say that X is projectively normal (for the given embedding) if S[X] is integrally closed.

VII.2.1. PROPOSITION. *Let* k *be a field and* A *a finitely generated* k-*algebra. If* X *is a connected normal subscheme of* $\mathbb{P}^r_A$, *then* :

1. $S' \bigoplus_{n\geq 0} \Gamma(X, \mathcal{O}_X(n))$ *is the integral closure of the domain* S[X].
2. *For* $d \in \mathbb{N}$ *large enough, we have* $S[X]_d = S'_d$.
3. *An arbitrary closed subscheme* $X \subset \mathbb{P}^r_A$ *is projectively normal if and only if* X *is normal and for every* $n \in \mathbb{N}$ *the map*

$$\Gamma(\mathbb{P}^r_k, \mathcal{O}_{\mathbb{P}^r_k}(n)) \to \Gamma(X, \mathcal{O}_X(n))$$

is surjective.

PROOF. Cf. [49]. □

VII.2.2. LEMMA. *Let* R *be a graded domain generated as an* R_0-*algebra by a finite number of elements of degree 1. Every homogeneous element of non-negative degree in* $Q^g_{R_+}(R)$ *is integral over* R.

PROOF. Let $\{f_0,\dots,f_n\} \subset R_1$ generate R as an R_0-algebra. Pick $f \in Q^g_{R_+}(R)$ homogeneous of positive degree d, then for some $n \in \mathbb{N}$, $R^n_+ f \subset R$. In particular $R^n_1 f \subset R_{n+d}$ and we obtain $R^n_1 f^2 \subset R^{n+d}_1 f \subset R^{n+2d}_1$. Consequently, $R^n_1 f^s \subset R$ for all $s \in \mathbb{N}$. In particular $f^n_0 f^s \subset R$ hence $R[f] \subset f^{-n}_0 R$. Since the latter is a Noetherian R-module, it follows that R[f] is finitely generated, hence f is integral over R. □

VII.2.3. COROLLARY. *In the foregoing* R_0 *is a field then so is* $(Q^g_{R_+}(R))_0 = \Gamma(\mathrm{Proj}(R), \tilde{R})$.

PROOF. Pick $0 \neq x \in Q^g_{R_+}(R)_0$ then x is integral over the field R_0 and contained in a domain i.e. x is invertible. □

VII.2.4. COROLLARY. *Let* R *be an affine graded domain over a field* k, *then* $Q^g_{R_+}(R)$ *is positively graded, unless* $\mathrm{Proj}(R) = \{0\}$.

PROOF. By assumption R is of the form $k[X_0,\dots,X_n]/I$, where I is a graded prime ideal of $k[X_0,\dots,X_n]$. Suppose we may find $0 \neq r \in Q^g_{R_+}(R)_{-n}$ with $n > 0$; since $R_n \neq 0$, we may pick a nonzero f in R_n and so $fr \in Q^g_{R_+}(R)_0$ is invertible in $Q^g_{R_+}(R)_0$ by Corollary VII.2.3., hence r is invertible in $Q^g_{R_+}(R)$. For arbitrary $r' \in Q^g_{R_+}(R)_n$ we may multiply r' by r^q for some q sufficiently large such that $r^q r' \in Q^g_{R_+}(R)_{-p}$ for some $p > 0$. Since $r^q r' \neq 0$, the foregoing argument yields that $r^q r'$ is invertible in $Q^g_{R_+}(R)$ and therefore r' is invertible in $Q^g_{R_+}(R)$. Consequently $Q^g_{R_+}(R) = Q^g_{R-P}(R)$ for all $P \in \mathrm{Proj}(R)$, meaning that $\mathrm{Proj}(R) = 0$. This situation having been excluded, the assertion follows. □

VII.2.5. COROLLARY. *If* R *is the homogeneous coordinate ring of a connected normal projective* k*-variety, then* $Q^g_{R_+}(R)$ *is the integral closure of* R.

VII.2.6. COROLLARY. *Let* R *be an affine graded domain over a field* k*, then* $Q^g_{R_+}(R)$ *is a graded Noetherian domain.*

PROOF. We have already established that $Q^g_{R_+}(R)$ is positively graded and that for large $d \geq d_0$, $(Q^g_{R_+}(R))_d = R_d$ (cf. Section VII.1.). By Serre's theorem, cf. [49], we know that the part of $Q^g_{R_+}(R)$ of degree lower than d_0 is a finitely generated R_0-module, hence Noetherian as well.

VII.2.7. LEMMA. *Let* A *be a* $\kappa^g_{R_+}$*-Azumaya algebra over* R*; write* $R' = Q^g_{R_+}(R)$. *For all* $P \in \mathrm{Proj}(R)$ we have that $Q^g_{R-P}(A)$ *is a graded Azumaya algebra over* $Q^g_{R-P}(R')$.

PROOF. Since R' is a graded integral extension of R it follows that $P = P' \cap R$ is in Proj(R) if P' is in Proj(R'). Indeed if P' contains R_+ then it also contains any $x \in h(R')_+$ because such an x satisfies a relation $x^n = r_1 x^{n-1} + \dots + r_n$ for some $r_i \in h(R_+)$, $i = 1,\dots,n$. Now $Q^g_{R'-P'}(A) = Q^g_{R'-P'}(R') \otimes_{Q^g_{R-P}(R)} Q^g_{R-P}(A)$. Since $Q^g_{R-P}(A)$ is a graded $Q^g_{R-P}(R)$-Azumaya algebra the assertion follows. □

At this point we recall the theory expounded in Section VI.3., in particular the definition of a graded reflexive Azumaya algebra over R (before Proposition VI.3.23.) and Theorem VI.3.24. describing the group $\beta^g(R)$. With this terminology :

VII.2.8. PROPOSITION. *Let* $X = \mathrm{Proj}(R)$ *be a connected regular projective* k-*variety of dimension at most two, then there is a bijective correspondence between locally separable sheaves of* $\mathcal{O}_X$-*algebras and the graded reflexive Azumaya algebras over* $\Gamma_*(\tilde{R})$.

PROOF. Since $Q^g_{R_+}(R)$ is Noetherian and integrally closed we have that, for every graded finitely generated torsion-free $Q^g_{R_+}(R)$-module M, $M \subset Q^g_{R_+}(M) \subset M^{**}$ (see Proposition VI.3.16.). So if M is reflexive then it is $\kappa^g_{R_+}$-closed. On the other hand, if $Q^g_{R_+}(M)$ is reflexive then $Q^g_{R_+}(M) = M^{**}$. If L is a locally separable sheaf of $\mathcal{O}_X$-algebras then $A = \Gamma_*(L)$ is reflexive and $Q^g_{R_+}(A \underset{R}{\otimes} A^0) = \mathrm{END}_R(A)$. Since the first is reflexive it follows that $Q^g_{R_+}(A \underset{R}{\otimes} A^0) = (A \underset{R}{\otimes} A^0)^{**}$ and therefore A is a graded reflexive Azumaya algebra in the sense of Section VI.3. Conversely if A is such a graded R-algebra then A is in particular finitely generated and torsion free, so it is certainly $\kappa^g_{R_+}$-closed. From the fact that $\Gamma_*(\tilde{R}) = Q^g_{R_+}(R)$ is Noetherian it follows that the graded $Q^g_{R-P}(R)$-modules $Q^g_{R-P}(A)^{**}$ and $Q^g_{R-P}(A)$ are isomorphic in degree zero. Consequently $Q^g_{R-P}(A)_0$ is a reflexive $Q^g_{R-P}(R)_0$-module. Then $Q^g_{R-P}(A)_0$ is a projective $Q^g_{R-P}(R)_0$-module, since $Q^g_{R-P}(R)_0$ is a regular ring of dimension at most two. This establishes that $Q^g_{R-P}(A)$ is a graded projective $Q^g_{R-P}(R)$-module, i.e. A will be a $\kappa^g_{R_+}$-quasiprojective R-module. Finally we arrive at the fact that $Q^g_{R_+}(A \underset{R}{\otimes} A^0)$ is a $\kappa^g_{R_+}$-finitely generated $\kappa^g_{R_+}$-closed and $\kappa^g_{R_+}$-quasiprojective R-module, but then $Q^g_{R_+}(A \underset{R}{\otimes} A^0)$ is reflexive, and we obtain :

$$Q^g_{R_+}(A \underset{R}{\otimes} A^0) = (A \underset{R}{\otimes} A^0)^{**} = \mathrm{END}_R(A)$$

i.e. A is a graded $\kappa^g_{R_+}$-Azumaya algebra. Now we may apply Theorem VI.1.17. and the result follows. □

The following theorem is now a direct consequence of the foregoing proposition :

VII.2.9. THEOREM. *Let* Proj(R) *be a connected regular projective* k-*variety of dimension at most two, then the canonical group morphism :*

$$\beta^g(\Gamma_*(\tilde{R})) \to \mathrm{Br}(\mathrm{Proj}(R),\tilde{R}),\ [L] \to [\tilde{L}],$$

is an isomorphism of abelian groups. □

In the non-regular case we are still able to give a complete description of Br(Proj(R)) in terms of graded relative Brauer groups if we restrict our attention to curves. So from hereon R will be a positively graded Noetherian domain generated by R_1 over the field $k = R_0$. In particular, the integral closure $\overline{R}$ of R is a finite R-module. It is well-known that $\overline{R}$ is a graded ring and that the conductor of $\overline{R}$ in R is a common graded ideal of R and $\overline{R}$. With these conventions :

VII.2.10. LEMMA. *Let* c *be the conductor of* $\overline{R}$ *in* R. *For any* $f \in h(R_+)$, *the following properties hold :*

1. $Q_f^g(\overline{R})$ *is the integral closure of* $Q_f^g(R)$.
2. *The conductor* d *of* $Q_f^g(\overline{R})$ *in* $Q_f^g(R)$ *is* $Q_f^g(c)$.

PROOF. 1. Obvious, because Q_f^g is obtained by inverting $\{1,f,f^2,\dots\}$.
2. Since $Q_f^g(c) = Q_f^g(R)c$ it is clear that $Q_f^g(c) \subset d$. Conversely, if $yf^{-n} \in d$ then $yf^{-n}x_i \in Q_f^g(R)$ for each $i = 1,\dots,m$, where $x_1,\dots,x_m$ is a set of homogeneous generators for $\overline{R}$ over R. Hence $yx_i \in Q_f^g(R)$, $i = 1,\dots,m$, and for large enough p, $f^p y x_i \in R$, $i = 1,\dots,m$. Consequently we have $f^p y\overline{R} \subset R$ i.e. $f^p y \in c$ or $yf^{-n} \in Q_f^g(c)$. □

VII.2.11. COROLLARY. Proj($\overline{R}$) *is the normalization of* Proj(R).

PROOF. Immediately from the local description of the normalization of a scheme in terms of an affine covering. □

VII.2.12. PROPOSITION. *For* $P \in \mathrm{Proj}(R)$, *the following statements are equivalent :*

1. *The conductor* c *of* $\overline{R}$ *in* P *is contained in* P.
2. $Q^g_{R-P}(R)_0 = R_{(P)}$ *is not integrally closed.*
3. $Q^g_{R-P}(R)$ *is not integrally closed.*

PROOF. 1 ⇒ 2. Pick $f \in R_1 - P$, then $Q^g_f(c)_0 \subset Q^g_f(P)$ and thus the conductor c' of $Q^g_f(\overline{R})_0$ in $Q^g_f(R)_0$ is contained in $Q^g_f(P)_0$. Obviously, $Q^g_{R-P}(R) = Q^g_{P^e}(Q^g_f(R))$ where $P^e = Q^g_f(P)$. Since $Q^g_f(R) = Q^g_f(R)_0[f,f^{-1}]$ we obtain the graded localization at $Q^g_f(P)$ by inverting the elements of $Q^g_f(R)_0 - Q^g_f(P)_0$, i.e. $Q^g_{R-P}(R)_0 = Q_{P^e}(Q^g_f(R)_0)$ and also it follows that $R_{(P)}$ is not integrally closed.

2 ⇒ 1. Again choose an $f \in R_1 - P$. Since $Q^g_f(R)_0$ is not integrally closed, $Q^g_f(c)_0 \subset Q^g_f(P)$ (as in the first part of this proof). However

$$Q^g_f(R) = Q^g_f(R)_0[f,f^{-1}]$$

then entails that $Q^g_f(P) \cap R = P$.

2 ⇔ 3. Take $x \in Q^g_{R-P}(\overline{R})_{-s}$, then we have a relation :

$$x^n + r_1x^{n-1} + \ldots + r_n = 0, \text{ with } r_i \in Q^g_{R-P}(R)_{is},$$

$i = 1,\ldots,n$. Select $f \in Q^g_{R-P}(R)_1$ such that $Q^g_{R-P}(R) = R_{(P)}[f,f^{-1}]$. We obtain a new relation :

$$(f^sx)^n + (f^sr_1)(f^sx)^{n-1} + \ldots + (f^{ns}r_n) = 0,$$

where now all the coefficients are in $R_{(P)}$; hence $f^sx \in R_{(P)}$ but then it follows that $x \in Q^g_{R-P}(R)$; i.e. $Q^g_{R-P}(R)$ is integrally closed. The converse implication may be derived in a similar way. □

If $X = \mathrm{Proj}(R)$ is a curve, then $\tilde{X} = \mathrm{Proj}(\overline{R})$ is its normalization. Let $\tilde{c}$ be the conductor sheaf $\mathrm{Ann}_X(\pi_*\mathcal{O}_{\tilde{X}},\mathcal{O}_X)$ where $\pi : \tilde{X} \to X$ is the canonical covering. Let V, resp. $\tilde{V}$, be the closed subscheme of X, resp. $\tilde{X}$, determined by c.

VII.2.13. LEMMA. $\mathrm{Pic}(\tilde{V}) = 0$.

PROOF. $\mathrm{Pic}(\tilde{V}) = H^1(\tilde{V},\mathcal{O}^*_{\tilde{V}})$ and $\dim \tilde{V} = 0$, since $\tilde{V} \cong \mathrm{Proj}(\bar{R}/c)$. It follows that $H^1(\tilde{V},\mathcal{O}^*_{\tilde{V}}) = 0$, cf. [49]. □

With assumptions as before we obtain the important theorem relating the Brauer groups of the schemes $X,\tilde{X}$ and $U,\tilde{V}$.

VII.2.14. THEOREM. *There is an exact sequence :*

$$0 \to \mathrm{Br}(X) \to \mathrm{Br}(\tilde{X}) \oplus \mathrm{Br}(V) \to \mathrm{Br}(\tilde{V}).$$

PROOF. Our assumptions imply that X may be covered by two open affines whose intersection is affine too, because X is separated. The same is true for the other schemes in the diagram :

$$\begin{array}{ccc} \tilde{X} & \xrightarrow{i_1} & \tilde{V} \\ \pi\downarrow & & \downarrow \\ X & \xrightarrow{i} & V \end{array}$$

For the étale topology, there is an exact sequence of sheaves on X, cf. [59] :

$$0 \to \mathbb{G}_{m,X} \to \pi_*\mathbb{G}_{m,\tilde{X}} \oplus i_*\mathbb{G}_{m,V} \to (\pi i_1)_*\mathbb{G}_{m,\tilde{V}} \to 0.$$

This yields a long exact sequence :

(*)
$$\ldots \to \mathrm{Pic}(\tilde{V}) \to H^2_{\acute{e}t}(X,\mathbb{G}_m) \to H^2_{\acute{e}t}(X,\mathbb{G}_m) \oplus H^2_{\acute{e}t}(V,\mathbb{G}_m) \to H^2_{\acute{e}t}(\tilde{V},\mathbb{G}_m).$$

Putting $\mathrm{Pic}(\tilde{V}) = 0$ and using the fact that in the situation we are considering, i.e. the schemes are covered by two nicely behaved affines, we have $H^2_{\acute{e}t}(-,\mathbb{G}_m)_t = \mathrm{Br}$ (the Brauer group is the torsion part of the second étale cohomology with values in the multiplicative group), and this reduces (*) to the desired exact sequence.

If one wants to avoid the use of étale cohomology and the deep results of [59], [39], an elementary proof may be given as follows (we give a sketch here). If S is an arbitrary positively graded ring let $\mathbb{P}(S)$, resp. $\mathbb{AZ}(S)$, be the category of locally projective sheaves of $\tilde{S}$-modules of finite type, resp. of Azumaya Algebras on Proj(S). The Cartesian diagram :

$$\begin{array}{ccc} R & \to & \overline{R} \\ \downarrow & & \downarrow \\ R/c & \to & \overline{R}/c \end{array}$$

defines in the natural way a commutative square :

$$(**) \qquad \begin{array}{ccc} \mathbb{P}(R) & \to & \mathbb{P}(\overline{R}) \\ \downarrow & & \downarrow \\ \mathbb{P}(R/c) & \to & \mathbb{P}(\overline{R}/c) \end{array}$$

Let $\mathbb{P}(\overline{R}) \underset{\mathbb{P}(\overline{R}/c)}{\times} \mathbb{P}(R/c)$ be the corresponding Cartesian product, then we may pass on to an affine covering and apply Milnor's theorem to derive in a rather straightforward way that the canonical functor

$$\mathbb{P}(\overline{R}) \underset{\mathbb{P}(\overline{R}/c)}{\times} \mathbb{P}(R/c) \to \mathbb{P}(R)$$

is in fact an isomorphism i.e. that (**) is again a Cartesian diagram. In a similar way one then checks the analogous statement for $\mathbb{AZ}$. Then, using Lemma VII.2.13. one may mimick the elementary proof given for the module theoretic analogue of the theorem. □

After this result we only need to calculate Br(V), Br($\tilde{V}$) explicitly in order to be able to say that we described Br(X) accurately in terms of Br($\tilde{X}$) which is easy to calculate (such statements are always to be understood from the point of view of relativity). Since X is a curve, V contains a finite number of points, $V = \{P_1,\ldots,P_n\}$ say. Note that we do not distinguish whether we view a prime P as an element of $V_+(c) \subset \mathrm{Proj}(R)$ or as an element of Proj(R/c). Since X is a curve, $\bigcap_{j\neq i} P_j \not\subset P_i$, so we always may choose an $f_i \in h(\bigcap_{j\neq i} P_j - P_i)_+$, $i = 1,\ldots,n$, so we obtain a set of homogeneous elements $f_1,\ldots,f_n \in h(R_+)$ such that $X_+(f_j) \cap V = \{P_j\}$. It follows that V is the disjoint union of the affines $\mathrm{Spec}((Q^g_{f_i}(R/c))_0)$, i.e. $\mathrm{Br}(V) = \bigoplus_{i=1}^n \mathrm{Br}((Q^g_{R-P}(R)/Q^g_{R-P}(c))_0)$ because

$$\mathrm{Spec}((Q^g_{f_i}(R/c))_0) = \mathrm{Spec}((Q^g_{R-P}(R)/Q^g_{R-P}(c))_0).$$

For any $P \in V$, we claim that the reduced ring $(Q^g_{R-P}(R)/\mathrm{rad}(Q^g_{R-P}(c)))_0$ is just the function field $\mathbb{k}_X(P)$. Because for any graded ring S we have that $(S/\mathrm{rad}(S))_0 = S_0/\mathrm{rad}(S_0)$, the ring considered is just :

$$R_{(P)}/\mathrm{rad}(Q^g_{R-P}(c))_0.$$

Now the primes in V are the only graded prime ideals of R (different from R_+) containing c; moreover the prime radical of a graded ideal of R is again a graded ideal of R, so this implies that $\mathrm{rad}(Q^g_{R-P}(c)) = Q^g_{R-P}(P)$ and this establishes our claim. Write $R_{red} = R/\mathrm{rad}(R)$. Taking into account the fact that $\mathrm{Br}(R) = \mathrm{Br}(R_{red})$ holds for any Noetherian ring, cf. [31], we have proved the following theorem :

VII.2.15. <u>THEOREM</u>. *Let* $X = \mathrm{Proj}(R)$ *be a connected projective curve. Let* $V, \tilde{V}$ *be the closed subschemes of* X *resp.* $\tilde{X}$ *determined by the conductor* c *of* $\overline{R}$ *in* R. *There is an exact sequence of group morphisms :*

$$0 \to \mathrm{Br}(X) \to \beta^g(\overline{R}) \oplus \Big(\bigoplus_{P\in V} \mathrm{Br}(\mathbb{k}_X(P))\Big) \to \bigoplus_{Q\in\tilde{V}} \mathrm{Br}(\mathbb{k}_{\tilde{X}}(Q)).$$

<u>PROOF</u>. Starting from the exact sequence in Theorem VII.2.14. it suffices to apply the preceding remarks to V as well as $\tilde{V}$. □

Some explicit examples are included as exercises.

CHAPTER VIII
EXERCISES

VIII.1. Chapter I.

1. Let I be a finitely generated left ideal of an arbitrary non-commutative ring, show that the filter $L(I) = \{L$ left ideal of $R;\ L \supset I^n$, some $n \in \mathbb{N}\}$ determines an idempotent kernel functor κ_I.

2. Give a proof for Proposition I.1.4.

3. If κ is an idempotent kernel functor having property T, show that $\kappa = V\{\kappa_I;$ I finitely generated $I \in L(\kappa)\}$.

4. Give an example of a Dedekind ring R and an ideal I of R such that κ_I cannot be obtained as a localization at a multiplicatively closed set i.e. by inverting some elements of R.

5. Calculate : $\mathrm{Pic}(\mathbb{Z}[\sqrt{2}])$, $\mathrm{Pic}(\mathbb{Z}[\sqrt{-5}])$, $\mathrm{Pic}(\mathbb{Z}[\sqrt{5}])$.
   Solve the general case $\mathrm{Pic}(\mathbb{Z}([\sqrt{d}])$, $d \in \mathbb{Z}$, using any book on elementary number theory.

6. Let R be any commutative domain, $a,b \in R$, and consider a quaternion-ring $H_{a,b}$ over R defined by adding symbols u,v to R subjected to the usual quaternion relations : $u^2 = a$, $v^2 = b$, $uv = -vu$. Check whether $H_{a,b}$ is an Azumaya algebra over R in case $\mathrm{char}(R) \neq 2$.

7. Let J be an ideal of the R-algebra A contained in the Jacobson radical J(A). Suppose that A/J is separable over R. An idempotent

$e \in A^e = A \otimes_R A^0$ is said to be an inertial idempotent if $m_A(e) = 1$ and the image of e in $(A/J)^e$ is a separability idempotent for A/J.

a) If A^e contains an idempotent f mapping to a separability idempotent of A/J then A contains an inertial idempotent.

b) If A is separable over R then $A \otimes_{Z(A)} A^0 = (A \otimes_R A^0)e_Z$ where e_Z is the separability idempotent for Z(A) over R.

c) If e is an inertial idempotent for a finitely generated separable R-algebra A then e is a separability idempotent for A.

8. Let A be an R-algebra, consider $J \subset J(A)$ as in 7. An inertial subalgebra S of A is an R-subalgebra which is R-separable and such that $S + J = A$.

a) Let S be an inertial subalgebra of A with respect to J; let I be an ideal contained in J such that A/I is a separable R-algebra, then S is an inertial subalgebra with respect to I.

b) Suppose J is a nilpotent ideal of the R-algebra A and let S be a finitely generated R-submodule of A which is also an inertial subalgebra with respect to J. Any inertial idempotent e of A^e may be written as $e = e_S + x + y$ where e_S is the image of a separately idempotent for S, $x \in i(S \otimes_R J^0)$, $y \in i(J \otimes_R A^0)$, i being the canonical map.

c) As in b) but supposing that $J^2 = 0$. If e is an inertial idempotent, put $B'(A,e) = \{a \in A, (a \otimes 1 - 1 \otimes a)e \text{ is in } i(J \otimes_R A^e)\}$. Show that B'(A,e) is conjugated to S and thus it is also an inertial subalgebra with respect to J.

d) Let R be such that J(R) is nilpotent and suppose that A is an R-algebra which is an R-progenerator such that A/J(A) is R-separable. Then A contains an inertial subalgebra.

e) Let R be a commutative Henselian ring and let A be an R-algebra which is finitely generated as an R-module. Prove that inertial subalgebras are conjugate.

Note. Exercises 7, 8 stem from [E1], they are not trivial and not even easy.

9. Calculate : $Br(\mathbb{R}[T,T^{-1}]) = \mathbb{Z}/2\mathbb{Z} \oplus \mathbb{Z}/2\mathbb{Z}$.

VIII.2. Chapter II.

10. Consider $M,N \in (R,\kappa)$-mod and let $f : M \to N$ be an R-linear map. Show that f is an isomorphism in (R,κ)-mod if and only if Ker f, Coker f are κ-torsion modules.

11. Prove that $Pic(R,S,\kappa) = Pic(j_S,\kappa)$, with terminology and notations as in Lemma II.18. and consequent.

12. Let $R \to R_1$ be a surjective ring homomorphism and let κ be an idempotent radical kernel functor on R-mod. Explain why $Pic(R_1,\kappa)$ makes sense, i.e. show that κ induces a radical κ_1 on R_1-mod.

13. Let R be a Krull domain an put $\kappa = \inf\{\kappa_{R-P};\ P \in X^{(1)}(R)\}$.
    a) Show that every reflexive R-module is κ-flat.
    b) What does it mean for R if $Pic(R,\kappa_{R-P}) = 0$, for some $P \in Spec(R)$, in particular $P \notin X^{(1)}(R)$?
    c) What does it mean for R if $Pic(R,\kappa) = 0$ for every κ ?

14. Give an example where $Pic(R,\kappa) \neq Pic(Q_\kappa(R))$.

15. Calculate $Pic(\mathbb{Z}[G])$ where G is :
    a) The cyclic group of four elements
    b) The "Vierer" group i.e. the non-cyclic group of four elements.

16. In the paper by E. Jespers, F. Van Oystaeyen, K. Zeeuwts, "Control Subgroups and Localization of Group Rings". Comm. in Algebra, cf. [E2] , the authors study reduction of certain problems to smaller group rings i.e. localization problems are reduced to similar problems over a suitable control subgroup associated to the torsion theory considered. Use this technique to descend $Pic(\mathbb{Z}[G],\kappa)$ to $Pic(\mathbb{Z}[G_\kappa],\kappa')$ where κ' is defined in $Z[G_\kappa]$-mod, G_κ the control subgroup associated to κ.

We have not checked all the details of this "exercise" yet ! Further related results are to be found in H. Bass, P. Murthy, cf. [13], i.e. concerning Picard groups of abelian group rings. We hope to come back to these matters in more detail, either in Part II for some non-commutative considerations, or elsewhere.

VIII.3. Chapter III.

17. a. Show that every Galois extension of $\mathbb{Z}[\sqrt{2}]$ is of the form $\mathbb{Z}[\sqrt{2}] \oplus \ldots \oplus \mathbb{Z}[\sqrt{2}]$ i.e. the only Galois extension of $\mathbb{Z}[\sqrt{2}]$ having only trivial idempotents in $\mathbb{Z}[\sqrt{2}]$ itself.

    b. What happens for $\mathbb{Z}[\sqrt{-5}]$?

    c. Study the general case : $\mathbb{Z}[\sqrt{d}]$, $d \in \mathbb{Z}$.
    Hint. Study ramification of primes and use your favorite book on number theory once more.

18. Let J be a nilpotent ideal of R, then $Br(R/J) = Br(R)$. This is almost equivalent to rewriting [31], the reader better consult that paper before trying to solve the exercise. Also one needs exercises 7 and 8.

19. Calculate arbitrary relative Brauer groups for a Noetherian Prüfer domain.
    Hint. Use Proposition III.3.13.

20. Let R be a commutative k-algebra, k a field. Let φ be an automorphism of k which extends to an automorphism of R, again denoted by φ. Consider the ring of twisted polynomials $A = R[X,\varphi]$, $\varphi(r)X = Xr$ for every $r \in R$. Suppose also that $R^{(\varphi)} = \{r \in R;\ \varphi(r) = r\}$ is such that R is finitely generated over $R^{(\varphi)}$.
    Show that the R-algebra endomorphisms of A are automorphisms if $R = R^{(\varphi)} \otimes_{k^{(\varphi)}} k$.

21. Find an example, or a class of examples, of relative Azumaya algebras which need not be Azumaya algebras.

Hint. 1. Consider reflexive algebras over a Krull domain e.g. endomorphism rings of certain reflexive modules.

2. Consider maximal orders over certain homogeneous coordinate rings of projective curves (non-singular).

3. Consider maximal orders over coordinate rings of certain affine curves with singularities.

4. Consider any finitely generated prime central R-algebra A. Put $\kappa_m = \inf\{\kappa_{R-p}$, P has maximal P.I. degree, $p = P \cap R\}$. Suppose moreover that R is a Noetherian domain. Show that $Q_{\kappa_m}(A)$ is a κ_m-Azumaya algebra over $Q_{\kappa_m}(R)$.

5. Consider crossed products over relative Galois extensions.

22. Give a complete proof of Theorem III.4.2. and calculate the special cases mentioned in the final paragraph of Chapter III.

23. Let R be a local Noetherian domain, with maximal ideal M. Let A be a central R-algebra which is a finitely generated R-module. Is A a κ-Azumaya algebra for some κ ?

VIII.4. Chapter IV.

24. Derive Mayer-Vietoris sequences for the Brauer groups of ringed spaces (similar techniques as the ones used in VI.4. may be introduced for sheaves and presheaves).

25. (R. Treger). Let S be a Galois covering of some local normal domain R with Galois group G and let $M \subset S$ be a reflexive ideal such that for every $\sigma \in G$ the two S-modules σM and M are isomorphic, then show that $End_R(M)$ is an Azumaya algebra. (Hint : use the fact that an R-algebra A is an Azumaya algebra iff it is central, projective of finite type and A_p is an Azumaya R_p-algebra for all $p \in Spec(R)$ with $htp = 1$).

26. Let R be a regular noetherian ring and I an ideal of R, then $U = X(I)$ is an affine open set in Spec(R) if and only if $ht(I) \leqslant 1$. Prove this.

Show also that if R is local moreover, then U is necessarily of the form X_f for some $f \in R$.

27. (D. Ferrand). Let X be an arbitrary scheme M an $\mathcal{O}_X$-Module. The relation $x \in \{\overline{y}\}$ for $x,y \in X$ induces an order on X and since $x \in \{\overline{y}\}$ implies that there is a canonical map $M_x \to M_y$, we thus obtain a projective system $\{M_x;\ x \in X\}$ indexed by x. Put $P(M) = \varprojlim_{x\in X} M_x$. Let A be a commutative ring and M an A-module.
We call M(A-) punctual if the canonical map $M \to P(\tilde{M})$ is an isomorphism.
   a) Show that if A is semilocal, then all A-modules are punctual.
   b) More generally, show that if A is arbitrary and $M \in$ A-mod has the property that the canonical map $M \to \prod_p M_p$ is injective for some <u>finite</u> set of prime ideals $p \in \mathrm{Spec}(A)$, then M is punctual. It follows that if M is an A-module such that $\mathrm{Ass}_f(M)$ is finite, then M is punctual.
   c) As a corollary, show that if A has a semilocal ring of fractions, then any flat A-module is punctual.
   d) Generalize the foregoing to arbitrary, not necessarily affine schemes. (This is highly non-trivial !)

28. Reconsider the situation of exercise 23. The argumentation following Theorem III.4.1. may be applied to show that A is the module of sections of some quasicoherent sheaf A over $Y = \mathrm{Spec}(R) - \{M\}$, if and only if $A = Q_M(A)$. How does this relate to A being a κ_M-Azumaya algebra ?

29. Let X be the affine plane with a finite number of points and lines removed, $X = \mathbb{A}_k^2 - \{f_i, P_j\}$. Then $\mathrm{Br}(X) = \mathrm{Br}(k)$.

30. Let X be a curve, show that $\mathrm{Br}(X_{reg}) = \bigcap_{x\in X_{reg}} \mathrm{Br}(\mathcal{O}_{X,x})$, where X_{reg} stands for the subscheme of X induced on the set of nonsingular points of X.

31. Let X be any curve. Let U be an open set in X such that $U \subset X_{reg}$. Show that $\mathrm{Br}(U) = \mathrm{Br}(\Gamma(U))$, where $\Gamma(U)$ is the coordinate ring of U.

32. Let X be an affine curve over a finite field. Establish that there is an exact sequence :

$$0 \to Br(X) \to \bigoplus_i (Q/\mathbb{Z})_i \to Q/\mathbb{Z} \to 0,$$

where the sum is over all points of Y - X where Y is a regular completion of X; cf. also E. Witt, Zerlegung reeller algebraischer Funktionen in Quadrate, Schiefkörper über rellem Funktionenkörper, J. Reine Angew. Math. 171, 1934, 4-11.

33. (Cf. [32]). Let X be an affine curve in $\mathbb{A}^n_k$, k = $\mathbb{C}$. Let s be the number of real components of X. Then $Br(\Gamma(X)) = (\mathbb{Z}/2\mathbb{Z})^s$.

34. Compute the following Brauer groups :
a) $Br(k[X,Y]/(X^2+Y^2-1))$
b) $Br(k[X,Y]/(X^4+Y^2(y^2-1))$.

VIII.5. CHAPTER V.

35. Let R be an integrally closed noetherian domain and let σ be the idempotent functor associated to the prime ideals of height 1. For any finite set of variables T, prove that Pic(R[T],σ) = Cl(R).

36. Let R be a Krull domain and $\sigma = \inf\{\kappa_{R-P};\ P \in X^{(1)}\}$. An R-lattice is divisorial if and only if it is torsion-free and σ-closed. Let A be a not necessarily commutative R-algebra and let M, N be R-lattices which are also A-modules. Show that $Hom_A(M,N)$ is divisorial if N is.

37. If M and N are torsion free modules over the Krull domain R, then their modified tensorproduct is defined to be $M \perp N = Q_\sigma(M \otimes_R N)$, with notations as before. Show that if N is a divisorial lattice then $M \perp N = M \otimes_R N$.

38. (Cf. [ ]). Let R be a Krull domain and consider the category A(R) whose objects are R-algebras A which are R-lattices with the property that the canonical map $A \perp A^{opp} \to End_R(A)$ is an isomorphism. As

in the Noetherian case, two objects $A,B \in A(R)$ are equivalent ($A \sim B$) if we may find divisorial R-lattices P,Q and an isomorphism $A \perp \mathrm{End}_R(P) = B \perp \mathrm{End}_R(Q)$. Show that the set of equivalence classes $\beta(R)$ of objects in $A(R)$ for this relation forms a group, with multiplication induced by the modified tensorproduct. Show also that a suitable modification of the techniques developed in III.4. and V.2. yields an exact sequence

$$0 \to \mathrm{Pic}(R) \to \mathrm{Cl}(R) \to \mathrm{Bcl}(R) \to \mathrm{Br}(R) \to \beta(R)$$

where Bcl(R) is defined as in V.2.

39. (Cf. [67]). Let R be a Krull domain and let $X^{(1)} \subset Y \subset \mathrm{Spec}(R)$. Define $I_Y(R)$ to consist of divisorial R-lattices M such that $M_p = I(p) \otimes_{R_p} R_p^n$ for all $p \in Y$, where $n = \mathrm{rk}_R M$ and I(p) is an invertible R_p-module.
   a) Prove that if M,N are divisorial lattices with M, $M \perp N \in I_Y(R)$, then $N \in I_Y(R)$.
   b) Give a torsion-theoretic interpretation of $I_Y(R)$.
   c) Give a geometric interpretation of $I_Y(R)$ if Y is open in the Zariski topology.
   d) Using the modified tensorproduct, it is clear how one may define a "Brauer group" of R based on $I_Y(R)$; denote it by $\mathrm{Br}_Y(R)$. Show that $\mathrm{Ker}(\mathrm{Br}(R) \to \prod_{p\in Y} \mathrm{Br}(R_p)) = \mathrm{Ker}(\mathrm{Br}(R) \to \mathrm{Br}_Y(R))$, all maps being canonical ones.

40. Let X be an integral noetherian scheme. If M is a coherent sheaf on X, one defines the dual of M to be the coherent sheaf $M^{\vee} = \mathcal{H}om(M, \mathcal{O}_X)$; we call M reflexive if the canonical map $M \to M^{\vee\vee}$ is an isomorphism.
   a) Prove that a coherent sheaf M on X is reflexive if and only if (locally) it can be included in an exact sequence

$$0 \to M \to M' \to M'' \to 0$$

   where M' is locally free and M'' is torsion free.

b) If X is normal, then this is equivalent to M being torsion free and for every $x \in X$ such that $\dim \mathcal{O}_{X,x} \geq 2$ we have depth $M_x \geq 2$. (Hint : use Serre's criterion "normal $\Leftrightarrow R_1 + S_2$", cf. [49].)

VIII.6. Chapter VI.

41. Show that the following properties are equivalent for graded modules M :

    a) M is finitely generated gr-projective

    b) M is finitely presented gr-flat.

42. Prove Proposition VI.1.7.

43. Let D be a graded commutative domain, $D = \bigoplus_{s \in S} D_s$, where S is a cancenllative torsionless monoid.
    For any monoid homomorphism $f : S \to G$ into a totally ordered additive group G, we may define a valuation $v : D^* \to G$ by putting $v(d_{s_1} + \ldots + d_{s_t}) = \inf\{f(s_i), i = 1,\ldots,t\}$; check that v is indeed a valuation.

44. Use the torsion theory (expounded in [66]) in R-gr to develop the graded version of the theory for relative Picard groups. Relate $\mathrm{Pic}^g(R,\kappa^g)$ and $\mathrm{Pic}(R,\kappa)$ in case κ is a kernel functor of finite type (e.g. if R is Noetherian) i.e. in case $Q^g_{\kappa^g} = Q_\kappa$ on R-gr. In this context it is important to note that a radical graded kernel functor (in the sense of [66]) $\kappa^g = \inf\{\kappa^g_{R-P}, P \in C(\kappa)\}$ also has the gr-radical property : $\kappa^g = \inf\{\kappa^g_{R-P}, P \in C^g(\kappa)\}$.

45. A positively graded gr-Dedekind ring D is of the form k[X] where k is a field.

46. Let A be a graded Azumaya algebra over the gr-Dedekind ring D which is not positively graded. Show that there is a bijective correspondence between $\mathrm{Spec}_g(A)$ and $\mathrm{Proj}(A_{\geq 0})$, where $A_{\geq 0} = \bigoplus_{n \geq 0} A_n$. Cf. [92].

47. Give an example of a graded R-algebra A such that $Z(A_0) \supsetneq (Z(A))_0$.

Show that, in case A is neither left nor right limited graded, $Z(A_{\geq 0}) = Z(A)_{\geq 0}$.

48. Put, $C = (\mathbb{C}[X,Y,Z]/(X^2+Y^2+Z^2))_{(X,Y,Z)}$,

$D = (\mathbb{R}[X,Y,Z]/(X^2+Y^2+Z^2))_{(X,Y,Z)}$.

Show that the canonical morphism Br C → Br Q(C) is monomorphic, whereas the canonical morphism Br(D) → Br(Q(D)) is not monomorphic. Define C^g, D^g in the similar way but localizing in the graded sense at the graded prime ideal (X,Y,Z), derive similar properties for $Br^g(C^g) \to Br^g(Q^g(C^g))$, $Br(C^g) \to Br(Q(C^g) = Br(Q(C))$.

49. (V.I. Danilov). Let X be a smooth homogeneous projective plane curve with homogeneous equation F(X,Y,Z) = 0. Let A be the affine coordinate ring of the cone on X, A = k[X,Y,Z]/(F). Put M = (x,y) x = X mod(F), y = Y mod(F). Then $Br(A_M) \to Br(\mathbb{k}(F))$ is monomorphic.

50. The reader interested in general graded ring theory might be willing to solve all exercises listed in [66].

51. Let R be a gr-field. Any proper graded ideal I in $S = R[X_1,\ldots,X_n]$, where the latter is graded by putting

$$S_m = \sum_{i+t_1+\ldots+t_n=m} R_i X_1^{t_1} \ldots X_n^{t_n},$$

has a zero $(x_1,\ldots,x_n)$ such that x_i, i = 1,...,n, is contained in some graded algebraic extension of R.

52. If V is a gr-valuation ring in the gr-field R then for every set of homogeneous elements $\{a_1,\ldots,a_n\} \in R - V$ there is an index i such that $a_i^{-1}a_j \in V$ for all j = 1,...,n.

53. Let S be a graded domain, let R be the gr-field of fractions of S and consider a gr-field extension T of R. Show that the integral closure of S in T is a graded ring and that it equals the integral closure of S in the field of fractions Q(T) of T. In particular : show that the integral closure of a graded domain in its field of fractions is again a graded domain.

54. Put $R = k[T,T^{-1}]$, $\deg T = d$, $K = k(T) = Q(R)$.
Consider $L = k(\alpha)$ where α satisfies the equation $\alpha^2 + T\alpha - T^3 = 0$. Show that one cannot find a gradation on R defined by putting $\deg T = t$, such that L is the field of fractions of the graded integral closure of R in L, cf. [86].

55. Prove the graded version of the commutator theorem, cf. [33], Theorem 4.3.

56. Let I be a nilpotent ideal of a graded commutative ring R. Show that $Br^g(R) \to Br^g(S/I)$ is an isomorphism. Hint : use the hint given in the proof of Lemma VI.4.9.

57. The integral closure $\overline{R}$ of a positively graded domain is again positively graded.

58. If a graded Azumaya algebra A over R is gr-split by a graded commutative ring S then there is a B, a graded Azumaya algebra over R which is gr-equivalent to A in $Br^g(R)$, such that S is a maximal (graded) commutative subring of B.

59. Provide examples (at least two of a different type) of discrete gr-valuation rings D which are not of the form $D_0[X,X^{-1}]$, X a variable.

60. Prove that the graded completion defined in Section VI.5. may indeed be calculated by the rule : $\hat{R}^g = \bigoplus_{n\in\mathbb{Z}} \hat{R}_n$, given in VI.5.7.

61. Prove that in general the uniformity M_I-gr is not metrizable.

62. Let I be a graded ideal of the graded ring R. Show that the following statements are equivalent :

a) Given $a_1,a_2,\ldots \in R_r$ such that $a_{i+1} - a_i \in (I^i)_r$ there exists an $a \in R_r$ such that $a - a_i \in (I^i)_r$ for all i.

b) Given $S \subset \mathbb{Z}$ a finite set together with $a_1,a_2,\ldots \in R_S$ such that $a_{i+1} - a_i \in (I^i)_S$, then there is an $a \in R_S$ such that $a - a_i \in (I^i)_S$ for all i, where $R_S = \bigoplus_{n\in S} R_n$, $(I^i)_S = \bigoplus_{n\in S} (I^i)_n$.

63. A gr-complete gr-local ring R is gr-Henselian.

Hint : first check that a gr-free finitely generated R-algebra, B say, is gr-$B\omega$-complete, where ω is the unique gr-maximal ideal of R. Therefore $B_r \cong \varprojlim_n B_r/(B\omega^n)_r$.

Since ω/ω^n is a nilideal of R/ω^n, the map $\mathrm{Id}(B/B\omega^n) \to \mathrm{Id}(B/B\omega)$ is a bijection. Since R/ω is a gr-field one may check that it is also gr-Henselian, so the foregoing implies that R/ω^n is gr-Henselian too. Consequently the graded R/ω^n-algebra $B/B\omega^n$ is gr-decomposed, use the decomposition to obtain that

$$(B/B\omega^n)_r \cong \bigoplus_i (Q^g_{B-P_i}(B)/\omega^n Q^g_{B-P_i}(B))_r,$$

where the P_i are the gr-maximal ideals of B lying over ω. Taking limits in the foregoing expression yields the result.

64. If G is an abelian group, give an elementary proof of :

$$H^2(G,U(S))/H^2_{gr}(G,U(S)) = H^2_{sym}(G,U(S))/H^2_{gr}(G,U(S)) \cap H^2_{sym}(G,U(S))$$

where S is a Galois extension of R with Galois group G.

65. If R is strongly graded, show that $\beta(R_0) = \beta^g(R)$.

66. If κ is a graded kernel functor on R-gr (cf. [66]), show that $\mathrm{Br}^g(R,\kappa) = \mathrm{Br}(R_0,\kappa_0)$ holds if R is strongly graded: κ_0 is the kernel functor induced on R_0-mod via the equivalence R-gr $\to$ R_0-mod (this equivalence does most of the work in the proof !) We have only hinted at the construction of $\mathrm{Br}^g(R,\kappa)$, the reader may check some of the basic properties for $\mathrm{Br}^g(R,\kappa)$ which are similar to those of $\mathrm{Br}(R,\underline{\kappa})$, where $\underline{\kappa}$ stands for the kernel functor induced by κ on R-mod by forgetting gradations.

VIII.7. Chapter VII.

67. Calculate : $\mathrm{Br}(\mathrm{Proj}(k[X,Y,Z])/(X^2+Y^2))) = \mathbb{Z}/2\mathbb{Z}$. Calculate Br(F) where F is your favorite projective curve.

68. Let R be the homogeneous coordinate ring of a connected normal projective k-variety. Show that $Q^g_{R_+}(R)$ is

a) positively graded,

b) the integral closure of R,

c) a graded Noetherian domain.

If k is not algebraically closed we may have that $(Q^g_{R_+}(R))_0 = 1$ is different from k; give an example.

VIII.8. Chapter VIII.

Some of the exercises are very long ones leading to some new applications or to further abstraction. An example of the first kind is : 16; for the second kind : 24, 44, 66. Some exercises are interrelated, for example 43 and 52 or rather the gr-valuation rings appearing in 52. An explicitation of this relation may be rewarding too i.e. in the case mentioned it is clear that the valuations mentioned in 43 extend the notion of gr-valuation ring, and the connection with the theory of gr-primes of rings becomes evident.

We did not include too much exercises of the type : calculate Pic(-,κ) or $Pic^g(-)$ or $Br^g(-)$ or Br(-,κ) of the ring (-).
The reader may easily exercise his ability in this kind of computing by considering his favorite rings and graded rings.
Inspiration for some exercises stems from the following papers :

E.1. IGRAHAM E.C., Inertial Subalgebras of Algebras over Commutative Rings, Trans. Amer. Math. Soc. 124, 1966, 77-93.

E.2. JESPERS E., VAN OYSTAEYEN F., ZEEUWTS K., Control Subgroups and Localization, Comm. in Algebra 9, 12, 1981, 1263-1285.

E.3. WITT E., Zerlegung reeller algebraischer Functionen in Quadrate Schiefkörper über reellem Funktionenkörper, J. Reine Angew. Math. 171, 1934, 4-11.

REFERENCES

[1] AMITSUR S.A., Homology Groups and Double Complexes for Arbitrary Fields, J. Math. Soc. Japan 14, 1962, 1-25.

[2] ANDERSON D., Graded Krull Domains, Comm. in Algebra 7, 1974, 79-106.

[3] ANDREOTTI A., BOMBIERI E., Supli omeomorfismi delle varieta algebriche, Ann. Scuola Norm. Sup. Pisa, 23, 1969, 430-450.

[4] ANDREOTTI A., NORGUET F., La convexité holomorphe dans l'espace des cycles d'une variété algébrique, Ann. Scuola Norm. Sup. Pisa, 21, 1967, 31-82.

[5] ARTIN M., Etale Coverings of Schemes over Hensel Rings, Amer. J. Math. 88, 1966, 915-934.

[6] ARTIN M., On the Joins of Hensel Rings, Adv. in Math. 7, 1971, 282-296.

[7] ARTIN M., Azumaya Algebras and Finite Dimensional Representations of Rings, J. of Algebra 11, 1969, 532-563.

[8] AUSLANDER M., The Brauer Group of a Ringed Space, J. of Algebra 4, 1966, 220-273.

[9] AUSLANDER M., GOLDMAN O., Maximal Orders, Trans. Amer. Math. Soc. 97, 1960, 1-24.

[10] AUSLANDER M., GOLDMAN O., The Brauer Group of a Commutative Ring, Trans. Amer. Math. Soc. 97, 1960, 367-409.

[11] AZUMAYA G., On Maximally Central Algebras, Nagoya Math. J. 2, 1951, 119-150.

[12] BASS H., Algebraic K-Theory, Benjamin, New York 1968.

[13] BASS H., MURTHY P., Grothendieck Groups and Picard Groups of Abelian Group Rings, Ann. of Math. 86, 1967, 16-73.

[14] BERGMAN G., On Jacobson Radicals of Graded Rings, preprint.

[15] BOURBAKI N., Algèbre Commutative, Eléments de Math. 27, 28, 30, 31, Hermann, Paris, 1961-66.

[16] CAENEPEEL S., Cohomological Interpretation of the Graded Brauer Group I, preprint, V.U. Brussels.

[17] CAENEPEEL S., Graded Complete and Graded Henselian Rings, preprint V.U. Brussels.

[18] CAENEPEEL S., VAN OYSTAEYEN F., Crossed Products over Graded Local Rings, in "Brauer Groups in Ring Theory and Algebraic Geometry", Lect. Notes in Math. 917, Springer Verlag, Berlin 1982.

[19] CARTAN M., EILENBERG S., Homological Algebra, Princeton University Press, Princeton 1956.

[20] CHASE S., HARRISON P., ROSENBERG A., Galois Theory and Cohomology of Commutative Rings, Mem. Amer. Math. Soc. 52, 1965, 1-19.

[21] CHASE S., ROSENBERG A., Amitsur Cohomology and the Brauer Group, Mem. Amer. Math. Soc. 52, 1965.

[22] CHILDS L., On Projective Modules and Automorphisms of Central Separable Algebras, Canad. J. Math. 21, 1969, 44-53.

[23] CHILDS L., Brauer Groups of Affine Rings, in "Ring Theory", Proceedings of the Oklahoma Conference, Lect. Notes in Pure and Appl. Math., Marcel Dekker, New York, 1974, 83-93.

[24] CHILDS L., Mayer-Vietoris Sequences and Brauer Groups of non-Normal Domains, Trans. Amer. Math. Soc. 196, 1974, 51-67.

[25] CHILDS L., GARFINKEL G., ORZECH M., The Brauer Group of Graded Azumaya Algebras, Trans. Amer. Math. Soc. 175, 1973, 299-326.

[26] CHILDS L., GARFINKEL G., ORZECH M., On the Brauer Group and Factoriality of Normal Domains, J. Pure Appl. Algebra 6, 1975, 111-123.

[27] CLABORN L., FOSSUM R., Generalization of the Notion of Class Group, Ill. J. Math. 12, 1968, 228-253.

[28] DADE E., Compounding Clifford's Theory, Ann. of Math. 2, 91, 1970, 236-270.

[29] DADE E., Group Graded Rings and Modules, Math. Z. 174, 3, 1980, 241-262.

[30] DE MEYER F., Projective Modules over Central Separable Algebras, Canad. J. Math. 21, 1969, 39-43.

[31] DE MEYER F., The Brauer Group of a Ring Modulo an Ideal, Rocky Mtn. J. of Math. 6, 1976, 191-198.

[32] DE MEYER F., The Brauer Group of Affine Curves, in "Brauer Groups", Evanston 1975", Lect. Notes in Math. 549, Springer-Verlag, Berlin 1976, 16-24.

[33] DE MEYER F., INGRAHAM E., Separable Algebras over Commutative Rings, Lect. Notes in Math. 181, Springer Verlag, Berlin 1971.

[34] DE MEYER F., KNUS M., The Brauer Group of a Real Curve, Proc. Amer. Math. Soc. 57, 1976, 227-232.

[35] FORMANEK E., Central Polynomials for Matrix Rings, J. of Algebra 23, 1972, 129-132.

[36] FOSSUM R., The Divisor Class Group of a Krull Domain, Ergebnisse der Math. 74, Springer Verlag, Berlin 1973.

[37] FOSSUM R., FOXBY H., The Category of Graded Modules, Math. Scand. 35, 1974, nr. 2.

[38] FRÖHLICH A., The Picard Group of Noncommutative Rings in Particular of Orders, Trans. Amer. Math. Soc. 180, 1973, 1-45.

[39] GABBER O., Some Theorems on Azumaya Algebras, in Groupe de Brauer Lect. Notes in Math. 844, Springer Verlag, Berlin 1981, 129-209.

[40] GABRIEL P., Des Catégories Abéliennes, Bull. Soc. Math. France 90, 1962, 323-448.

[41] GABRIEL P., POPESCU N., Caractérisation des Catégories abéliennes avec générateurs et limites inductives exactes, C.R. Ac. Sci. Paris, 258, 1964, 4188-4190.

[42] GODEMENT R., Théorie des faisceaux, Hermann, Paris 1961.

[43] GOLAN J., Localization in Noncommutative Rings, Pure and Appl. Math., Marcell Dekker, New York 1975.

[44] GOLDMAN O., Rings and Modules of Quotients, J. of Algebra 13, 1969, 10-47.

[45] GROTHENDIECK A., Local Cohomology, Lect. Notes in Math. 1, Springer Verlag, Berlin 1967.

[46] GROTHENDIECK A., Sur quelques points d'algèbre homologique, Tohoku Math. J. 9, 1957, 119-221.

[47] GROTHENDIECK A., Le groupe de Brauer I, II, III, in "Dix Exposés sur la Cohomologie des Schémas, North Holland, Amsterdam 1968.

[48] GROTHENDIECK A., ARTIN M., VERDIER J-L., Théorie des topos et cohomologie étale des schémas, Lect. Notes in Math. 269, 270, 305, Springer Verlag, Berlin 1972-73.

[49] HARTSHORNE R., Algbraic Geometry, Springer Verlag, Berlin 1977.

[50] HOOBLER R., Br'(X) = Br(x) if X is the union of two affines, preprint 1981.

[51] ISCHEBECK F., Zur Picar Gruppe eines Produktes, Math. Z. 139, 1974, 141-157.

[52] JACOBSON N., Lectures in Abstract Algebra, vol. III, Van Nostrand 1964.

[53] KANZAKI R., On Generalized Crossed Product and Brauer Group, Osaka J. Math. 5, 1968, 175-188.

[54] KAPLANSKI I., Commutative Rings, Allyn and Bacon, Boston 1970.

[55] KATZ V., The Brauer Group of a Regular Local Ring, Ph.D. thesis, Brandeis University 1968.

[56] KNUS M.A., OJANGUREN M., Théorie de la descente et d'Algèbres d' Azumaya, Lect. Notes in Math. 389, Springer Verlag, Berlin 1974.

[57] KNUS M.A., OJANGUREN M., A Mayer-Vietoris sequence for the Brauer Group, J. Pure and Appl. Algebra 5, 1974, 345-360.

[58] MILNER J., The Brauer Group of a Rational Surface, Inv. Math. 11, 1970, 304-307.

[59] MILNER J., Etale Cohomology, Princeton University Press, Princeton, New Jersey 1980.

[60] MILNOR J., Introduction to Algbraic K-Theory, Princeton University Press, Princeton, New Jersey 1971.

[61] MACLANE S., Categories for the Working Mathematician, Springer Verlag, Berlin 1971.

[62] MORI Y., On the Integral Closure of an Integral Domain, Bull. Kyoto Univ. Ser. B, 7, 1955, 19-30.

[63] MURDOCH D., VAN OYSTAEYEN F., Noncommutative Localization and Sheaves, J. of Algebra 35, 1975, 500-515.

[64] NAGATA M., Local Rings, Inter Science Tracts in Pure and Appl. Math. 13, New York 1962.

[65] NASTASESCU C., VAN OYSTAEYEN F., Graded and Filtered Rings and Modules, Lect. Notes in Math. 758, Springer Verlag, Berlin 1979.

[66] NASTASESCU C., VAN OYSTAEYEN F., Graded Ring Theory, Library of Math., North-Holland, Amsterdam 1982.

[67] ORZECH M., Brauer Groups and Class Groups for a Krull Domain, in "Brauer Groups in Ring Theory and Algebraic Geometry", Lect. Notes in Math. 917, Springer Verlag, Berlin 1982.

[68] ORZECH M., SMALL C., The Brauer Group of a Commutative Ring, Lect. Notes in Pure and Appl. Math. 11, Marcel Dekker, New York 1975.

[69] ORZECH M., VERSCHOREN A., Some Remarks on Brauer Groups of Krull Domains, Lect. Notes in Math. 917, Springer Verlag, Berlin 1982, 91-96.

[70] PAREIGIS B., Non-additive ring and module theory IV : The Brauer Group of a Symmetric Monoidal Category, in "Brauer Groups", Evanston 1975, Lect. Notes in Math. 549, Springer Verlag, Berlin 1975.

[71] PROCESI C., Rings with Polynomial Identities, Marcel Dekker, New York 1973.

[72] RAYNAUD M., Anneaux Locaux Henseliens, Lect. Notes in Math. 169, Springer Verlag, Berlin 1970.

[73] REINER I., Maximal Orders, Academic Press, New York, 1975.

[74] ROSENBERG A., ZELINSKY D., On Amitsur's Complex, Trans. Amer. Math. Soc. 97, 1960, 327-356.

[75] SAMUEL P., Lectures on U.F.D., Tata Inst., vol. 30, Bombay 1964.

[76] SAMUEL P., Anneaux Gradués Factoriels et Modules Reflexifs, Bull. Soc. Math. France 92, 1964, 237-249.

[77] SCHILLING O., The Theory of Valuations, Amer. Math. Soc. Surveys 4, 1950.

[78] SERRE J-P., Cohomologie Galoisienne, Lect. Notes in Math. 11, Springer Verlag, Berlin 1964.

[79] SERRE J-P., Faisceaux Algébriques Cohérents, Ann. of Math. 61, 1955, 197-278.

[80] SONN J., Class Groups and Brauer Groups, Israel J. Math. 34, 1979, 97-106.

[81] STENSTRÖM B., Rings of Quotients, Springer Verlag, Berlin 1975.

[82] TAMONE G., Su une generalizzazione della nozione di incollamente, Reud. Sc. fis. mat. e. nat. Lincei 65, 1978, 107-112.

[83] TRAVERSO C., Seminormality and Picard Groups, Ann. Scuola Norm. Sup. Pisa 24, 1970, 585-595.

[84] TREGER R., Reflexive Modules, J. of Algbra 54, 1978, 444-466.

[85] VAN GEEL J., Primes and Value Functions, Ph.D. Thesis, U.I.A. Antwerp 1980.

[86] VAN OYSTAEYEN F., VAN GEEL J., On Graded Fields, Indag. Math. 43, 1981, 273-286.

[87] VAN OYSTAEYEN F., On Graded Rings and Modules of Quotients, Comm. in Algebra 6, 1978, 1923-1959.

[88] VAN OYSTAEYEN F., Prime Spectra in Non-commutative Algebra, Lect. Notes in Math. 444, Springer Verlag, Berlin 1975.

[89] VAN OYSTAEYEN F., Compatibility of Kernel Functors and Localization Functors, Bull. Soc. Math. Belg. XVIII, 1976, 131-137.

[90] VAN OYSTAEYEN F., On Brauer Groups of Arithmetically Graded Rings, Comm. in Algebra 9, 19, 1981, 1873-1892.

[91] VAN OYSTAEYEN F., Crossed Products over Arithmetically Graded Rings, J. of Algebra, to appear.

[92] VAN OYSTAEYEN F., Graded Azumaya Algebras and Brauer Groups, Proceedings Ring Theory, U.I.A. 1980, Lect. Notes in Math. 825, Springer Verlag, Berlin 1980.

[93] VAN OYSTAEYEN F., Generalized Rees Rings, J. of Algebra, to appear.

[94] VAN OYSTAEYEN F., VERSCHOREN A., Reflectors and Localization. Application to Sheaf Theory, Lect. Notes in Pure and Appl. Math. 41, M. Dekker, New York 1979.

[95] VAN OYSTAEYEN F., VERSCHOREN A., Noncommutative Algebraic Geometry, an Introduction, Lect. Notes in Math. 887, Springer Verlag, Berlin 1981.

[96] VAN OYSTAEYEN F., VERSCHOREN A., The Brauer Group of a Projective Variety, Israel J. of Math. 42, 1982, 37-59.

[97] VAN OYSTAEYEN F., VERSCHOREN A., The Brauer Group of a Projective Curve, Israel J. of Math., to appear.

[98] VERSCHOREN A., On the Picard Group of a Grothendieck Category, Comm. in Algebra 8, 1980, 1169-1194.

[99] VERSCHOREN A., Relative Seminormalization, Comm. in Algebra, to appear.

[100] VERSCHOREN A., Mayer-Vietoris Sequences for Brauer Groups of Graded Rings, Comm. in Algebra.

[101] VERSCHOREN A., The Brauer Group of a Quasi-Affine Scheme, Lect. Notes in Math. 917, Springer Verlag, Berlin 1982, 260-279.

[102] VILLAMAYOR O., ZELINSKY D., Brauer Groups and Amitsur Cohomology for General Commutative Ring Extensions, J. Pure Appl. Algebra 10, 1977, 198-202.

[103] YUAN S., Reflexive Modules and Algebra Class Groups over Noetherian Integrally Closed Domains, J. of Algebra 32, 1974, 405-417.

[104] ZARISKI O., SAMUEL P., Commutative Algebra, I, II, Van Nostrand.

[105] ZELINSKY D., Long Exact Sequences and the Brauer Group, Evanston 1975, Lect. Notes in Math. 549, Springer Verlag 1976, 63-70.

[106] ZELINSKY D., Brauer Groups, Lect. Notes in Pure Appl. Math. 26, M. Dekker, New York 1977.

Additional References.

Occasionally we have referred to results in papers which are not really in the vein of this book e.g. a result which is purely non-commutative or a result which has only an illustrative function. We list them here separately, for completeness' sake.

[107] BRAUN A., Affine P.I. Rings and Their Generalizations, J. Algebra 58, 1979, 481-494.

[108] CAUCHON G., Les T-anneaux et les anneaux à identités polynomiales Noethériens, Ph.D. Thesis, Paris XI, Orsay 1977.

[109] CHAMARIE M., Anneaux de Krull non-commutatifs, Ph.D. Thesis, Univ. Claude Bernard, Lyon I, 1981.

[110] HORROCKS G., Vector Bundles on the Punctured Spectrum of a Local Ring, Proc. London Math. Soc. (3) 14, 1964, 689-713.

[111] LAM T., Serre's conjecture, Lecture Notes in Math. 635, Springer Verlag, Berlin 1978.

[112] LAMBEK J., Torsion Theories, Additive Semantics and Rings of Quotients, Lect. Notes in Math. 177, Springer Verlag, Berlin 1977.

[113] LE BRUYN L., VAN OYSTAEYEN F., Generalized Rees Rings and Relative Maximal Orders satisfying Polynomial Identities, J. of Algebra, to appear.

[114] ROWEN L., On Rings with Central Polynomials, J. of Algebra 31, 1974, 393-426.

[115] VAN OYSTAEYEN F., On Orders over Graded Krull Domains, Osaka J. of Math., to appear.

[116] VAN ROOIJ A.C., Non-Archimedean Functional Analysis, Marcel Dekker Inc., New York, 1978.

INDEX

Y

Z